ANTHROPOLOGY AND MODERN LIFE

FRANZ BOAS

With an Introduction by
Ruth Bunzel

Dover Publications, Inc.
New York

This Dover edition, first published in 1986, is an unabridged republication of the 1962 edition published by W.W. Norton and Company, Inc., New York, who published the first edition in 1928.

Manufactured in the United States of America

Dover Publications, Inc., 31 East 2nd Street, Mineola, N.Y. 11501

Library of Congress Cataloging-in-Publication Data

Boas, Franz, 1858–1942.

Anthropology and modern life.

Reprint. Originally published: New York : Norton, 1962.

Bibliography: p.

1. Anthropology. 2. Race. 3. Social problems. I. Title.

[GN27.B6 1986] 306 86-19904

ISBN 0-486-25245-0 (pbk.)

CONTENTS

INTRODUCTION

"The American Anthropological Association repudiates statements now appearing in the United States that Negroes are biologically and in innate mental ability inferior to whites, and reaffirms the fact that there is no scientifically established evidence to justify the exclusion of any race from the rights guaranteed by the Constitution of the United States. The basic principles of equality of opportunity and equality before the law are compatible with all that is known about human biology. All races possess the abilities needed to participate fully in the democratic way of life and in modern technological civilization."

—Passed at the Annual Meeting of the Council of Fellows of the American Anthropological Association, November 17, 1961.

ON NOVEMBER 17, 1961, the Council of Fellows of the American Anthropological Association meeting at Philadelphia, the cradle of American democracy, passed this resolution, thus once more providing scientific support for those fighters for equality and brotherhood for whom democracy is a moral issue. At this moment of history when the specter of racism is once more walking abroad, it is especially fortunate and appropriate to have reissued in a popular edition the definitive statement on race and culture by the man who more than anyone else was responsible for providing the conceptual framework and scientific underpinnings for the anthropological position on this important contemporary problem. Franz Boas wrote *Anthropology and Modern Life* as a declaration of faith after more than thirty years of research in the field of race and culture. An earlier publication on the same theme was translated into German (*Kultur und Rasse,* Leipzig, 1914) and was eventually honored by a prominent place in the Nazi *auto-da-fé.*

When Boas first turned to anthropology in the closing decades of the nineteenth century, "ethnography" consisted largely of unsystematic observations of primitives by untrained observers and travelers, while "ethnology" consisted mainly of speculations on the history of civilization, with little reference to observed facts. Both approaches to the science of man were equally unrelated to the problems of modern life. So long as "savages" were regarded as a different species or an inferior and undeveloped branch of the human race, little could be learned from them, and the study of their strange customs had a purely antiquarian and collector's interest. Boas, however, early recognized the broader implications of anthropological studies. Writing in 1889 he said, "Investigations [of the different forms of family structure] show that emotional reactions which we feel as natural are in reality culturally determined. It is not easy for us to understand that the emotional relation between father and son should be different from the one to which we are accustomed, but knowledge of the life of people with a social organization different from ours brings about situations in which conflicts or mutual obligations arise of a character quite opposed to those we are accustomed to and that run counter to what we consider 'natural' emotional reactions to those to whom we are related by blood. The data of ethnology prove that not only our knowledge, but also our emotions are the result of the forms of our social life and of the history of the people to whom we belong. If we desire to understand the development of human culture we must try to free ourselves of these shackles. . . . We must lay aside many points of view that seem to us self-evident, because in early times they were not self-evident. It is impossible to determine *a priori* those parts of our mental life that are common to mankind as a whole and those due to the culture in which we live. A knowledge of the data of ethnology enables us to attain this insight. Therefore it enables us also to view our own civilization objectively."

One of the first controversies of the many that filled Boas' turbulent life was over the arrangement of museum collections, Boas staunchly defending his geographical and tribal classification against upholders of the more traditional arrangements by types of artifacts. He felt that one of the functions of a museum was to "educate and entertain" and that ethnological collections should be presented so as to illustrate ways of life rather than scientific typologies. His principles won out in all American museums (except the United States National Museum) as well as in many European museums. This was one of the many ways in which Boas sought to use anthropology to free men's minds of the yoke of traditional patterns of thought by confronting audiences with different and coherent styles of life.

Boas was educated in the tradition of liberal romanticism that produced Carl Schurz and the philosophical anarchists of the nineteenth century. He was the essential protestant; he valued autonomy above all things and respected the unique potentialities of each individual. He believed that man was a rational animal and could, with persistent effort, emancipate himself from superstition and irrationality and lead a sane and reasonable life in a good society—although he was fully aware that humanity had a long way to go to achieve this goal. This partly explains his unalterable opposition to Freud and psychoanalysis with its essentially tragic view of life and its acceptance of irrationality as an essential part of the human condition. During the last years of his life (he died during World War II) a deep depression overwhelmed him as he watched the rising tides of hatred and war. But although age and illness made him feel helpless, his faith in man never wavered. "If I were young I would *do* something," he said to a colleague who had remarked how difficult it must be for their students growing up in the midst of the Depression and under the threat of war. Always the activist!

For Boas, "doing something" always meant using his science in the cause of man. His object was the enlightenment of mankind through anthropology. He was a tireless lecturer, although he disliked public appearances and partial paralysis made speaking difficult for him. He was an indefatigable contributor to scientific journals and mass media, and a constant writer of "letters to the editor." As a teacher his influence was inestimable. He established anthropology as an academic discipline in America. Alexander F. Chamberlain, his student at Clark University, won the first doctorate in anthropology to be granted by an American university, and for more than forty years almost every anthropologist in America came directly or indirectly under Boas' influence. Among his students in the early days at Columbia were such distinguished anthropologists as Alfred L. Kroeber, Robert Lowie, Alexander Goldenweiser, Edward Sapir, Clark Wissler, Paul Radin, and Leslie Spier. During the twenties Ruth Benedict, Margaret Mead, Melville J. Herskovits and Otto Klineberg were all Boas students, as well as a host of less well-known scholars who set up departments and conducted research in all parts of the world. As a teacher Boas was a stern taskmaster; he made no concessions to ignorance. He gave students no reading lists or other aids; he opened his course in Biometrics with the statement, "I assume that you all know the calculus. If not, you will learn it." In his seminar he assigned books in Dutch or Portuguese; no student would dare to say to Boas, "I don't read Dutch." Somehow or other the student learned to cope. Boas rarely suggested subjects for dissertations; a student who had been studying anthropology for two years and had found no problem he wished to pursue was not worth bothering with. He would discuss general problems with students, but would not criticize or look at unfinished work. His criticisms were terse—"You have entirely missed the point"—and he almost never praised. One had to be tough, independent, and dedicated to survive. He was a formidable teacher and a formidable man. Yet, in spite of his apparent aloofness he was deeply concerned about his students, their lives and their careers, but generally in terms of what *he* thought was good for them. Although he valued autonomy, he was frequently high-handed. He arranged field trips and wangled jobs for students without consulting them and was deeply hurt if they refused to accept his arrangements. But he never wavered in his loyalty to students, however much he might disapprove of them. And his students, on their part, though some of them quarreled bitterly with him on theoretical and personal grounds, never lost their respect and loyalty. An *esprit de corps* united the group that shared the struggle to establish their science and communicate their ideas. It would be hard to duplicate today the ties that bound student to teacher and student to fellow sudent.

One of the areas in which Boas felt enlightenment was needed was in the problem of race. In the early paper already quoted he was pointing out the need to distinguish between those characteristics of a people which were biological and inherited and those which were acquired as part of that people's culture. This problem continued to occupy him throughout his life; it provides the unifying theme of *Anthropology and Modern Life.* Whenever he was examining the distribution of physical types of man, or national characteristics, or crime, or the rates of growth and maturation of children, he endeavored in carefully designed researches to separate man's culturally acquired characteristics from his innate endowment. Calling on history as his witness, he always insisted that the burden of proof was on those who would attribute differences to biological causes. Boas was trained in the natural sciences; what he carried over to his anthropological studies from his training in physics was not a specific method, for he realized early in his career that the methods of one discipline could not be applied to another and that the formulations of a social science must be of a different order from those of a laboratory science. He brought to anthropology rigorous standards of proof, a critical skepticism toward all generalizations, and the physicist's unwillingness to accept any generalization or explanation as anything more than a useful hypothesis until it had been clearly demonstrated that no other explanation was possible. This aspect of Boas' theoretic approach especially irked those of his colleagues who would have liked more facile generalization and who regarded Boas' standards of proof as a "methodological strait jacket."

In the field of physical anthropology he was a great innovator; he was interested only in the study of living people. The study of fossils and skeletal materials, which constituted a large part of the physical anthropology of the nineteenth century, did not interest him. He was dissatisfied with current definitions of race based on the selection of extreme forms as "pure" types, or the equally unsatisfactory definitions based on crude statistical "averages." He substituted populations, localized in space and time, for those vague entities, "races," as the units of study, thus foreshadowing contemporary trends in genetics. His observations on the instability of human types ("Changes in the Bodily Form of the Descendants of Immigrants," 1911) struck a body blow to theories of the immutability of racial characteristics. His conclusions aroused storms of criticism but were later fully corroborated. His studies of the growth of children had far-reaching results; not only did he introduce the concept of physiological as distinct from chronological age, with its influence on pediatrics and education, but his studies of children in different socio-economic backgrounds and especially his observations on the retardation of children in orphanages were instrumental in altering child-care programs and in the adoption of the foster-home plan.

In his emphasis on family lines, rather than race, as the mechanism of inheritance, he was establishing the scientific basis of individualism. Equality of races did not mean equality of individuals. Each individual human being is unique, the product of his own particular heredity, shared only by an identical twin, and of his life experience, including his culture. In a truly democratic society each individual, regardless of color, class, or sex is entitled to equal participation in the rewards of his culture, and the fullest development of his unique potentialities. Boas made his declaration of human rights in the name of science.

When Boas first visited the Eskimo he was confronted with the paradox of the unity and variety of human cultures—*plus ça change plus c'est la même chose.* Of the Eskimo he wrote: "After a long and intimate intercourse with the Eskimo, it was with feelings of sorrow and regret that I parted from my Arctic friends. I had seen that they enjoyed life, and a hard life, as we do; that nature is also beautiful to them; that feelings of friendship also root in the Eskimo heart; that, although the character of their life is so rude as compared to civilized life, the Eskimo is a man as we are; that his feelings, his virtues and his shortcomings are based in human nature, like ours."[1]

These two aspects of cultural anthropology were always present in his thinking and writing—the unity of man as a species, the universality of the basic pattern of his culture—the human biogram, as it came to be called—and human ingenuity in finding solutions to the problems of living in the various situations in which the accidents of time and history had placed him.

But Boas was no "cultural relativist" in the sense of thinking that there were no ethical absolutes. Eating one's neighbor is not a desirable or acceptable practice merely because the Eskimos do it from need and the Papuans from religious convictions. Such practices serve a function within the particular cultural settings in which they are found. The anthropologist must bring to the study of these phenomena the same detachment with which the biologist observes the predatory habits of tigers—who are not so predatory as the common stereotype would make them. But because anthropologists are studying human beings, and because we are involved with mankind and, in a deep sense, are our brothers' keepers, this detachment is hard to achieve without confusing moral sensibilities. *We* have not had to live with the daily prospect of starvation; *we* have not been taught to believe that the earth must be fertilized with human blood if it is to bear. We can afford to value each human life.

One of the popular misconceptions about Boas was that he was an anti-evolutionist. True, he did oppose the ethnocentric nineteenth-century version of cultural evolution—that mankind had evolved in a uniform series of stages from "savagery" to mid-Victorian England, and that all existing forms of culture were to be evaluated in terms of their similarity or dissimilarity to this most highly evolved culture.

---

[1] Quoted by Melville J. Herskovits in *Franz Boas: The Science of Man in the Making,* 1953.

But he believed, as must all who look at the long record of man's life on this planet, in cultural evolution. It was the method and the ethnocentric bias that he sought to correct. He believed not only in evolution, but in progress—specifically in two fields of human activity; in the growth of knowledge with its corollary of technology and man's increased control of his environment, and in man's growing control of his aggression which has enabled him to live at peace with ever larger groups of his fellows. Boas did not have available to him the great mass of material on primate behavior now extant which documents the devices in the animal world for maintaining peace within the group and between groups. He shared the nineteenth-century "tooth and claw" view of the animal world and visualized early man as living in a state of constant conflict. But he was right in recognizing the constant trend toward integrating larger and larger groups that was not only the result but the necessary condition of the advance in technology. In 1928 he saw that the inevitable next step was the integration of all mankind into one fellowship, since the interdependence of nations was making national rivalry untenable. Boas died in 1942, before the bomb fell on Hiroshima and the development of man's capacity to destroy himself made integration into one social system within which warfare was interdicted the very basis of survival.

Anthropology and Modern Life and its predecessor, *The Mind of Primitive Man,* are unpretentious books; they are written without jargon or pedantry; descriptive and illustrative material is cut to a minimum. But they are among the books which have changed men's minds. If some of the ideas developed in them now seem self-evident it is because they have become part of our thinking in the course of the more than thirty years since they were written. But old ways of thinking die hard and lingering deaths. It was necessary in 1961 for the American Anthropological Association to reaffirm its stand on racial equality. There are many who feel that only social systems that resemble ours are deserving of support; that the riches with which nature has blessed this country should be shared only with those who share our views. As we face the emerging nations of Africa and Asia we must take a long look backward at man's history on this earth and a long look forward to the next step in his evolution.

Columbia University
New York
January 15, 1962

RUTH BUNZEL

CHAPTER I

WHAT IS ANTHROPOLOGY?

ANTHROPOLOGY is often considered a collection of curious facts, telling about the peculiar appearance of exotic people and describing their strange customs and beliefs. It is looked upon as an entertaining diversion, apparently without any bearing upon the conduct of life of civilized communities.

This opinion is mistaken. More than that, I hope to demonstrate that a clear understanding of the principles of anthropology illuminates the social processes of our own times and may show us, if we are ready to listen to its teachings, what to do and what to avoid.

To prove my thesis I must explain briefly what anthropologists are trying to do.

It might appear that the domain of anthropology, of "the science of man," is preoccupied by a whole array of sciences. The anthropologist who studies bodily form is confronted by the anatomist who has spent centuries in researches on the gross form and minute structure of the human body. The physiologist and the psychologist devote themselves to inquiries into the functioning of body and mind. Is there, then, any justification for the anthropologist to claim that he can add to our fund of knowledge?

There is a difference between the work of the anthropologist and that of the anatomist, physiologist, and psychologist. They deal primarily with the typical form and function of the human body and mind. Minor differences such as appear in any series of individuals are either disregarded or considered as peculiarities without particular significance for the type, although sometimes suggestive of its rise from lower forms. The interest centers always in the individual as a type, and in the significance of his appearance and functions from a morphological, physiological or psychological point of view.

To the anthropologist, on the contrary, the individual appears important only as a member of a racial or a social group. The distribution and range of differences between individuals, and the characteristics as determined by the group to which each individual belongs are the phenomena to be investigated. The distribution of anatomical features, of physiological functions and of mental reactions are the subject matter of anthropological studies.

It might be said that anthropology is not a single science, for the anthropologist presupposes a knowledge of individual anatomy, physiology and psychology, and applies this knowledge to groups. Every one of these sciences may be and is being studied from an anthropological point of view.

The group, not the individual, is always the primary concern of the anthropologist. We may investigate a racial or social group in regard to the distribution of size of body as measured by weight and stature. The individual is important only as a member of the group, for we are interested in the factors that determine the distribution of forms or functions in the group. The physiologist may study the effect of strenuous exercise upon the function of the heart. The anthropologist accepts these data and investigates a group in which the general conditions of life make for strenuous exercise. He is interested in their effect upon the distribution of form, function and behavior among the individuals composing the group or upon the group as a whole.

The individual develops and acts as a member of a racial or a social group. His bodily form is determined by his ancestry and by the conditions under which he lives. The functions of the body, while controlled by bodily build, depend upon external conditions. If the people live by choice or necessity on an exclusive meat diet, their bodily functions will differ from those of other groups of the same build that live on a purely vegetable diet; or, conversely, different racial groups that are nourished in the same way may show a certain parallelism in physiological behavior.

Many examples can be given showing that people of essentially the same descent behave differently in different types of social setting. The mental reactions of the Indians of the western plateaus, a people of simple culture, differ from those of the ancient Mexicans, a people of the same race, but of more complex organization. The European peasants differ from the inhabitants of large cities; the American-born descendants of immigrants differ from their European ancestors; the Norse Viking from the Norwegian farmer in the northwestern States; the Roman republican from his degenerate descendants of the imperial period; the Russian peasant before the present revolution from the same peasant after the revolution.

The phenomena of anatomy, physiology and psychology are amenable to an individual, nonanthropological treatment, because it seems theoretically possible to isolate the individual and to formulate the problems of the variation of form and function in such a way that the social or racial factor is apparently excluded. This is quite impossible in all basically social phenomena, such as economic life, social organization of a group, religious ideas and art.

The psychologist may try to investigate the mental processes of artistic creation. Although the processes may be fundamentally the same everywhere, the very act of creation implies that we are not dealing with the artist alone as a creator but also with his reaction to the culture in which he lives and that of his fellows to the work he has created.

The economist who tries to unravel economic processes *must* operate with the social group, not with individuals. The same may be said of the student of social organization. It is possible to treat social organization from a purely formal point of view, to demonstrate by careful analysis the fundamental concepts underlying it. For the anthropologist this is the starting point for a consideration of the dynamic effects of such organization as manifested in the life of the individual and of the group.

The student of linguistics may investigate the "norm" of linguistic expression at a given time and the mechanical processes that give rise to phonetic changes; the psychological attitude expressed in language; and the conditions that bring about changes of meaning. The anthropologist is more deeply interested in the social aspect of the linguistic phenomenon, in language as a means of communication and in the interrelation between language and culture.

In short, when discussing the reactions of the individual to his fellows we are compelled to concentrate our attention upon the society in which he lives. We cannot treat the individual as an isolated unit. He must be studied in his social setting, and the question is relevant whether generalizations are possible by which a functional relation between generalized social data and the form and expression of individual life can be discovered; in other words, whether any generally valid laws exist that govern the life of society.

A scientific inquiry of this type is concerned only with the interrelations between the observed phenomena, in the same way as physics and chemistry are interested in the forms of equilibrium and movement of matter, as they appear to our senses. The question of the usefulness of the knowledge gained is entirely irrelevant. The interest of the physicist and chemist centers in the development of a complete understanding of the intricacies of the outer world. A discovery has value only from the point of view of shedding new light upon the general problems of these sciences. The applicability of experience to technical problems does not concern the physicist. What may be of greatest value in our practical life does not need to be of any interest to him, and what is of no value in our daily occupations may to him be of fundamental value. The only valuation of discoveries that can be admitted by pure science is their significance in the solution of general abstract problems.

While this standpoint of pure science is applicable also to social phenomena, it is easily recognized that these concern our own selves much more immediately, for almost every anthropological problem touches our most intimate life.

The course of development of a group of children depends upon their racial descent, the economic condition of their parents and their general well-being. A knowledge of the interaction of these factors may give us the power to control growth and to secure the best possible conditions of life for the group. All vital and social statistics are so intimately related to policies to be adopted or to be discarded that it is not quite easy to see that the interest in our problems, when considered from a purely scientific point of view, is not related to the practical values that we ascribe to the results.

It is the object of the following pages to discuss problems of modern life in the light of the results of anthropological studies carried on from a purely analytical point of view.

For this purpose it will be necessary to gain clarity in regard to two fundamental concepts: race and stability of culture. These will be discussed in their proper places.

CHAPTER II

THE PROBLEM OF RACE

In the present cultural conditions of mankind we observe, or observed at least until very recent time, a cleavage of cultural forms according to racial types. The contrast between European and East Asiatic civilizations was striking, until the Japanese began to introduce European patterns. Still greater appeared the contrasts between Europeans, native Australians, African Negroes and American Indians. It is, therefore, but natural that much thought has been given to the problem of the interrelation between race and culture. Even in Europe are found striking cultural differences between North Europeans and people of the Mediterranean, between West and East Europeans, and these are correlated with differences in physical appearance. This explains why numberless books and essays have been and are being written based on the assumption that each race has its own mental character determining its cultural or social behavior. In America particularly, fears are being expressed of the effects of intermixture of races, of a modification or deterioration of national character on account of the influx of new types into the population of our country, and policies of controlling the growth of the population are being proposed and laws based on these assumptions have been enacted.

In Melanesia the conflict of races finds expression in another way. In cases of intermarriages between a White man and a native woman the widow is liable to lose both the property left by her husband and the control of her children, and she is compelled, even if well educated, either to starve or to marry a native and to resume native life. This has happened even when the husband willed his property to his wife.

In South Africa the economic needs of natives and Whites have created sharp conflicts. A law was passed reserving certain districts exclusively for Whites, others exclusively for natives. The immediate result of this action has been that the natives were driven out by force from the White reservations, while the Whites who had settled in native reservations refused to go. The general policy of the Boers has been an attempt to suppress and exploit the native population.

The differences of cultural outlook and of bodily appearance have given rise to antagonisms that are rationalized as due to instinctive racial antipathies.

There is little clarity in regard to the term "race." We know only populations and we have to determine in how far population (or local race) and race are identical or distinct. When we speak of racial characteristics we mean those traits that are determined by heredity in each race and in which all members of the race participate. Comparing the color of skin, eyes and hair of Swedes and Negroes, slight pigmentation is a hereditary racial characteristic of the Swede, deep pigmentation of the Negro. The straight or wavy hair of the Swede, the frizzly hair of the Negro, the narrowness and elevation of the nose among the Swedes, its width and flatness among the Negroes, all these are hereditary racial traits because practically all the Swedes have the one group of characteristics, all the Negroes the other.

In other respects it is not so easy to define racial traits. Anatomists cannot with certainty differentiate between the brains of a Swede and of a Negro. The brains of individuals of each group vary so much in form that it is often difficult to say, if we have no other criteria, whether a certain brain belongs to a Swede or to a Negro.

The nearer two populations are related the more traits they will have in common. A knowledge of all the bodily traits of a particular individual from Denmark does not enable us to identify him as a Dane. If he is tall, blond, blue-eyed, long-headed and so on he might as well be a Swede. We also find individuals of the same bodily form in Germany, in France and we may even find them in Italy. Identification of an individual as a member of a definite population (or local race) is not possible.

Whenever these conditions prevail, we cannot

speak of racial heredity. In a strict sense the identification of a population as a race would require that *all* the members of the population partake of certain traits,—such as the hair, pigmentation and nose form of the Negro, as compared to the corresponding features among the North European. When only some members of each population have such distinguishing traits, while others are, in regard to their outer appearance or functioning, alike, then these traits are no longer true racial characteristics. Their significance is the less, the greater the number of individuals of each population that in regard to the feature in question may be matched. North Italians are roundheaded, Scandinavians long-headed. Still, so many different forms are represented in either series, and other bodily forms are so much alike that it would be impossible to claim that an individual selected at random *must* be a North Italian or a Scandinavian. Extreme forms in which the local characteristics are most pronounced might be identified with a fair degree of probability, but intermediate forms might belong to either group. The bodily traits of the two groups are not racial characteristics in the strict sense of the term. Although it is possible to describe the most common types of these groups by certain metric and descriptive traits, not all the members of the groups conform to them.

The bodily forms of Italians may serve as an example. The two most strongly contrasting types in Italy are the Piemontese and the Sardinians. We have records of the head forms, stature and hair color of these two groups. If I should assign, according to these three traits, individuals belonging to two identical populations entirely by chance to the one or the other I should err 125 times in 1,000 attempts. If I should have to decide whether they are Piemontese or Sardinians I should err 43 times in 1,000 attempts. Notwithstanding the great differences between the two groups the certainty of assignment is only one third of that of a chance assignment.

We are easily misled by general impressions. Most of the Swedes are blond, blue-eyed, tall and long-headed. This causes us to formulate in our minds the ideal of a Swede and we forget the variations that occur in Scandinavia. If we talk of a Sicilian we think of a swarthy, short person, with dark eyes and dark hair. Individuals differing from this type are not in our mind when we think of a "typical" Sicilian. The more uniform a people the more strongly are we impressed by the "type." Every country impresses us as inhabited by a certain type, the traits of which are determined by the most frequently occurring forms. This, however, does not tell us anything in regard to its hereditary composition and the range of its variations. The "type" is formed quite subjectively on the basis of our everyday experience.

We must also remember that the "type" is more or less an abstraction. The characteristic traits are found rarely combined in one and the same individual, although their frequency in the mass of the population induces us to imagine a typical individual in which all these traits appear combined.

The subjective value of the "type" appears also from the following consideration. Suppose a Swede, from a region in which blondness, blue eyes, tall stature prevail in almost the whole population, should visit Scotland and express his experiences naïvely. He would say that there are many individuals of Swedish type, but that besides this another type inhabits the country, of dark complexion, dark hair and eyes, but tall and long-headed. The population would seem to represent two types, not that biologically the proof would have been given of race mixture; it would merely be an expression due to earlier experiences. The unfamiliar type stands out as something new and the inclination prevails to consider the new type as racially distinct. Conversely, a Scotchman who visits Sweden would be struck by the similarity between most Swedes and the blond Scotch, and he would say that there is a very large number of the blond Scotch with whom he is familiar, without reaching the conclusion that his own type is mixed.

We speak of racial types in a similar way. When we see American Indians we recognize some as looking like Asiatics, others like East Europeans, still others are said to be of a Jewish cast. We classify the variety of forms according to our previous experiences and we are inclined to consider the divergent forms that are well established in our consciousness as pure types, particularly if they appear as extreme forms.

Thus the North European blond and the Armenian with his high nose and his remarkably high head which, when seen in profile, rises abruptly without a backward bulge, from the nape of the neck, appear as pure types.

Biologically speaking, this is an unjustifiable assumption. Extreme forms are not necessarily pure racial types. We do not know how much their descendants may vary among themselves and what their ancestry may have been. Even if it were shown that the extreme types were of homogeneous descent, this would not prove that the intermediate types might not be equally homogeneous.

It is well to remember that heredity means the transmission of anatomical and functional characteristics from ancestor to offspring. A population consists of many family lines whose descent from common ancestors cannot be proved.

The children of each couple represent the hereditarily transmitted qualities of their ancestors. Such a group of brothers and sisters is called a fraternity.

Not all the members of a fraternity are alike. They scatter around a certain middle value. If the typical distribution of forms in all the groups of brothers and sisters that constitute the population were alike, then we could talk of racial heredity, for each fraternity would represent the racial characteristics. We cannot speak of racial heredity if the fraternities are different, so that the distribution of forms in one family is different from that found in another one.

In this case the fraternities represent distinctive hereditary family lines. Actually in all the known populations the single family lines as represented by fraternities show a considerable amount of variation which indicates that the hereditary characteristics of the families are not the same, a result that may be expected whenever the ancestors have distinct or separable heritable characteristics. In addition to this we may observe that a fraternity found in one race may be duplicated by another one in another race; in other words, that the hereditary characteristics found in one race may not belong to it exclusively, but may belong also to other races.

This may be illustrated by an extreme case. If I wish to know "the type" of the New Yorker, I may not pick out any one particular family and claim that it is a good representative of the type. I might happen to select a family of pure English descent; and I might happen to strike an Irish, Italian, Jewish, German, Armenian or Negro family. All these types are so different and, if inbred, continue their types so consistently that none of them can possibly be taken as a representative New Yorker. Conditions in France are similar. I cannot select at random a French family and consider its members as typical of France. They may be blond Northwest Europeans, darker Central Europeans or of Mediterranean type. In New York as well as in France the family lines are so diverse that there is no racial unity and no racial heredity.

Matters are different in old, inbred communities.

If a number of families have intermarried for centuries without appreciable addition of foreign blood they will all be closely related and the same ancestral traits will appear in all the families. Brothers and sisters in any one family may be quite unlike among themselves, but all the family lines will have considerable likeness. It is much more feasible to obtain an impression of the general character of the population by examining a single family than in the preceding cases, and a few families would give us a good picture of the whole group. Conditions of this type prevail among the landowners in small European villages. They are found in the high nobility of Europe and also among some isolated tribes. The Eskimos of North Greenland, for instance, have been isolated for centuries. Their number can never have exceeded a few hundred. There are no rigid rules prescribing marriages between relatives, so that we may expect that unions were largely dictated by chance. The ancestors of the tribe were presumably a small number of families who happened to settle there and whose blood flows in the veins of all the members of the present generation. The people all bear a considerable likeness, but unfortunately we do not know in how far the family lines are alike.

We have information of this kind from one of the isolated Tennessee valleys in which people have intermarried among themselves for a century. The family lines in this community are very much alike.

In cases of this kind it does not matter whether

the ancestry is homogeneous or belongs to quite distinct races. As long as there is continued inbreeding the family lines will become alike. The differences of racial descent will rather appear in the differences between brothers and sisters, some of whom will lean towards one of the ancestral strains, others to the other. The distribution of different racial forms in all the various families will be the more the same, the longer the inbreeding without selection continues. We have a few examples of this kind. The Bastaards of South Africa, largely an old mixture of Dutch and Hottentot, and the Chippewa of eastern Canada, descendants of French and Indians, the mixed blood of Kisar, one of the islands of the Malay archipelago, descendants of Dutch and Malay, are inbred communities. Accordingly, the family lines among them are quite similar, while the brothers and sisters in each family differ strongly among themselves.

In modern society, particularly in cities, conditions are not favorable to inbreeding. The larger the area inhabited by a people, the denser and the more mobile the population, the less are the families inbred and the more may we expect very diverse types of family lines.

The truth of this statement may readily be demonstrated. Notwithstanding the apparent homogeneity of the Swedish nation, there are many different family lines represented. Many are "typical" blond Swedes, but in other families dark hair and brown eyes are hereditary. The range of hereditary forms is considerable.

It has been stated before that many individuals of Swedish type may be duplicated in neighboring countries. The same is true of family lines. It would not be difficult to find in Denmark, Germany, Holland or northern France families that might apparently just as well be Swedes; or in Sweden families that might as well be French or German.

This may be interpreted in one of two ways. It may be that the Swedish, German, Dutch, and northern French types are each of homogeneous ancestry but so variable that similar lines occur in all the groups; or the variations may be due to an intermingling of fundamentally different racial types, each of which is quite stable.

If we assume the former alternative we must say that the hereditary characteristics are not "racially" determined, but belong to family lines that occur in all these local groups. In this case the term "racial heredity" loses its meaning. We can speak solely of "heredity in family lines."

We may also assume that the population has originated through a mixture of distinct types. We have seen that our concept of types is based on subjective experience. On account of the preponderance of "typical" Swedes we are inclined to consider all those of different type as not belonging to the racial type, as foreign admixtures. There is a somewhat distinct type in Sweden in the old mining districts which were first worked by Walloons and it is more than probable that the greater darkness of complexion in this region is due to the influx of Walloon blood. We are very ready to explain every deviation from a type in this way. In many cases this is undoubtedly correct, for intermingling of distinct types of people has been going on for thousands of years; but we do not know to what extent a type may vary when no admixture of foreign blood has occurred. The experience of animal breeders proves that even with intensive inbreeding of pure stock there always remains a considerable amount of variation between individuals. We have no evidence to show to what extent variations of this kind might develop in a pure human race and it is not probable that satisfactory evidence will ever be forthcoming, because we have no pure races.

Even with the most intense amount of inbreeding and the most uniform characteristics of ancestors we must always expect a certain amount of variation of family lines, because the heritable characteristics are separable. Certain heritable forms may occur in one group of offspring, others in another. Uniformity could result only if all the traits of the ancestors were absolute units, unable to split up, a condition that does not occur in man.

To give an example: skin color may depend upon peculiar heritable characteristics in such a way that if the structure of the fertilized ovum varies in one direction pigmentation may be darker than if it varies in another direction. Then the members of the fraternity developing from these ova would vary in heritable skin color and the family lines established by them would differ, because the heritable character has been separated into distinct lines.

The history of the human races, as far as we can follow it, shows us mankind constantly on the move; people from eastern Asia migrating to Europe; those of western and central Asia invading southern Asia; North Europeans sweeping over Mediterranean countries; Central Africans extending their territories over almost the whole of South Africa; people from Alaska spreading to northern Mexico or vice versa; South Americans settling almost over the whole eastern part of the continent here and there; the Malay extending their migrations westward to Madagascar and eastward far over the Pacific Ocean—in short, from earliest times on we have a picture of continued movements, and with it of mixtures of diverse peoples.

It may well be that the lack of clean-cut geographical and biological lines between populations of different areas, even between the principal races of man is entirely due to these circumstances. The conditions are quite like those found in the animal world. Local races of remote districts may readily be recognized, but in many cases they are united by intermediate forms.

The assumption that each population consists of a mixture of racial types has led to the attempt to analyse it and to discover its component racial elements. In populations as similar as those of Europe, and without an intimate knowledge of the degree of morphologic stability of traits and of the detailed laws of heredity, types can be segregated only according to a purely subjective evaluation of traits. The effects of everyday experience in the establishment of types has been pointed out before. In the numerous attempts at such analysis pigmentation, form of hair, head, nose, and face, bodily build have been utilized, but no proof has ever been given that these may be genetically valid types and that the population is actually derived from such artificially constructed pure types. Even blue eyes, an apparently genetically fixed character, may have developed independently among various types due to the effect of domestication of man, as it has developed in many species of domesticated animals. In modern, mixed populations derived from fundamentally distinct races, like the Eurasians, Mulattoes, Zambos or American Mestizos we know the component elements and their influences can be studied in the family lines of the mixed population. If, conversely, we were required to reconstruct from the mixed population the unknown distinct types from which it is derived, we might be entirely misled in regard to their characteristic features. The establishment of "pure ancestral races" by means of analysis of populations is a venturous undertaking.

We have seen that on account of the lack of sharp distinctions between neighboring populations it happens that apparently identical family lines occur in both, and that an individual in one may resemble in bodily form an individual in another. Notwithstanding their resemblance it can be demonstrated that they are not by any means genetically equivalent, for when we compare their children they will be found to revert more or less to the type of the population to which the parents belong. To give an example: the Bohemians have, on the average, round heads, the Swedes long heads. Nevertheless it is possible to find among both populations parents that have the same head forms. The selected group among the Swedes will naturally be more round-headed than the average Swede, and the selected Bohemians will be more long-headed than the average Bohemian. The children of the selected group of Swedes are found to be more long-headed than their parents, those of the selected group of Bohemians more short-headed than their parents.

The cause of this is not difficult to understand. If we pick out short-headed individuals among the Swedes, short-headedness may be an individual non-hereditary trait. Furthermore the general run of their relatives will be similar to the long-headed Swedish type and since the form of the offspring depends not only upon the parent, but also upon the characteristics of his whole family line, at least of his four grandparents, a reversion to the general population may be expected. The same is true among the Bohemians.

We must conclude that individuals of the same

bodily appearance, if sprung from populations of distinct type, are genetically not necessarily the same. For this reason it is quite unjustifiable to select from a population a certain type and claim that it is identical with the corresponding type of another population. Each individual must be studied as a member of the group from which he has sprung. We may not assume that the round-headed or brunette individuals in Denmark are identical with the corresponding forms from Switzerland. Even if no anatomical differences between two series of such individuals are discernible they represent genetically distinctive strains. Identity can occur in exceptional individuals only.

If we were to select a group of tall, blond Sicilians, men and women, who marry among themselves, we must expect that their offspring in later generations will revert more or less to the Sicilian type, and, conversely, if we select a group of brunette, brown-eyed Swedes, their offspring will revert more or less to the blond, blue-eyed Swedish type.

We have spoken so far only of the hereditary conditions of stable races. We imply by the term racial heredity that the composition of succeeding generations is identical. When one generation dies, the next one is assumed to represent the same type of population. This can be true only if random matings, due to chance only, occur in each generation. If in the first generation there was a random selection of mates the same condition must prevail in the following generations. Any preferential mating, any selective change brought about by differential mortality or fertility, or by migration, must modify the genetic composition of the group.

For these reasons none of our modern populations is stable from a hereditary point of view. The heterogeneous family lines in a population that has originated through migration will gradually become more homogeneous, if the descendants continue to reside in the same spot. In our cities and mixed farming communities, on account of changes in selective mating, constant changes in the hereditary composition are going on, even after immigration has ceased. Local inbreeding produces local types; avoidance of marriages between near relatives favors increasing likeness of all the family lines constituting the population; favored or prescribed cousin marriages which are customary among many tribes establish separate family types and increase in this sense the heterogeneity of the population.

Another question presents itself. We have considered only the hereditary stability of genetic lines. We must ask ourselves also whether environmental conditions exert an influence over races.

It is quite obvious that the forms of lower organisms are subject to environmental influences. Plants taken from low altitudes to high mountains develop short stems; leaves of semi-aquatic plants growing under water have a form differing from that of their subaërial leaves. Cultivated plants transform their stamens into petals. Plants may be dwarfed or stimulated in their growth by appropriate treatment. Each plant is so organized that it develops a certain form under given environmental conditions. Microörganisms differ so much in different environmental settings that it is often difficult to establish their specific identity.

The question arises whether the same kind of variability occurs in higher organisms. The general impression is that their forms are determined by heredity, not by environment. The young of a greyhound is a greyhound, that of a shorthorn a shorthorn, that of a Norway rat a Norway rat. The child of a European is European in type, that of a Chinaman of Mongolic type, that of an African Negro a Negro.

Nevertheless detailed study shows that the form and size of the body are not entirely shaped by heredity. Records of stature that date back to the middle of the past century show that in almost all countries of Europe the average statures have increased by more than an inch. It is true, this is not a satisfactory proof of an actual change, because improvement in public health has changed the composition of the populations, and although it is not likely that this should be the cause of an increase in stature, it is conceivable. A better proof is found in the change of stature among descendants of Europeans who settle in America. In this case it has been shown that in many nationalities the children are taller than their own parents, presumably on account of more favorable conditions of life.

It has also been observed that the forms of the body are influenced by occupation. The hand of a person who has to do heavy manual labor differs from that of a musician who develops the independence of all the muscles of his hand. The proportions and forms of the limbs are influenced by habitual posture and use. The legs of the oriental who squats flat on the ground are somewhat modified by this habit.

Other modifications cannot be explained by better nutrition or by the use of the muscles. Forms of the head and face are not quite stable, but are in some way influenced by the environment in which the people live, so that after a migration into a new environment the child will not be quite like the parent.

All the observed changes are slight and do not modify the essential character of the hereditary forms. Still they are not negligible. We do not know how great the modifications may be that ultimately result from such changes, nor have we any evidence that the changes would persist if the people were taken back to their old environment. Although a Negro will never become a European, it is not impossible that some of the minor differences between European populations may be due to environment rather than to heredity.

So far we have discussed solely the anatomical forms of races with a view of gaining a clearer understanding of what we mean by the term race. It may be well to repeat the principal result of our discussion.

We have found that the term "racial heredity" is strictly applicable only when all the individuals of a race participate in certain anatomical features. In each race taken as a whole the family lines differ appreciably in their hereditary traits. The distribution of family lines is such that a considerable number of lines similar or even identical in one or many respects occur in contiguous territories. The vague impression of "types," abstracted from our everyday experience, does not prove that these are biologically distinct races, and the inference that various populations are composed of individuals belonging to various races is subjectively intelligible, objectively unproved. It is particularly not admissible to identify types apparently identical that occur in populations of different composition. Each individual can be understood only as a member of his group.

These considerations seem necessary, because they clear up the vagueness of the term "race" as usually applied. When we speak of heredity we are ordinarily concerned with family lines, not with races. The hereditary traits of families constituting the most homogeneous population differ very much among themselves and they are not sharply set off from neighboring populations that may give a quite distinctive impression.

The relation of racial types may be looked at in another way. It may be granted that in closely related types the identification of an individual as a member of each type cannot be made with any degree of certainty. Nevertheless the distribution of individuals and of family lines in the various races differs. When we select among the Europeans a group with large brains, their frequency will be relatively high, while among the Negroes the frequency of occurrence of the corresponding group will be low. If, for instance, there are 50 per cent of a European population who have a brain weight of more than, let us say, 1,500 grams, there may be only 20 per cent of Negroes of the same class. Therefore 30 per cent of the large-brained Europeans cannot be matched by any corresponding group of Negroes.

It is justifiable to compare races from this point of view, as long as we avoid an application of our results to individuals.

On general biological grounds it is important to know whether any one of the human races is, in regard to form or function, further removed from the ancestral animal form than another, whether the races can be arranged in an ascending series. Although we do not know the ancestral form with any degree of certainty, some of its characteristics can be inferred by a comparison of the anatomical forms of man and of the apes. Single traits can be brought into ascending series in which the racial forms differ more and more from animal forms, but the arrangement is a different one for each independent trait.

The ancestral form had a flat nose. Bushmen, Negroes and Australians have flat, broad noses. Mongoloids, Europeans and particularly Armenians have narrow, prominent noses. They are in this sense farthest removed from the animal forms.

Apes have narrow lips. The lips of the Whites are thin, those of many Mongoloid types are fuller. The Negroes have the thickest, most excessively "human" lips.

The hair coat of apes is moderately strong. Among human races the Australians, Europeans and a few scattered tribes among other races have the amplest body hair; Mongols have the least.

Similar remarks may be made in regard to the forms of the foot, of the spinal column, of the proportions of the limbs. The order of the degree to which human races differ from animals is not the same in regard to these traits.

Particular stress has been laid on the size of the brain, which also differs in various races. Setting aside the pygmy Bushmen and other very small races, the Negroid races have smaller brains than the Mongoloids, and these in general smaller ones than the Europeans, although some Mongoloid types, like the Eskimo, exceed in size of the brain many European groups.

The brain in each race is very variable in size and the "overlapping" of individuals in the races is marked. It is not possible to identify an individual as a Negro or White according to the size and form of the brain, but serially the Negro brain is less extremely human than that of the White.

We are apt to identify the size of the brain with its functioning. This is true to a limited extent only. Among the higher mammals the proportionate size of the brain is larger in animals that have greater intelligence; but size alone is not an adequate criterion. Complexity of structure is much more important than mere size. Some birds have brains much larger proportionately than those of the higher mammals without evidencing superior intelligence.

The size of the brain is measured by its weight which does not depend upon the nerve cells and fibers alone, but includes a large amount of material that is not directly relevant for the functioning of the central nervous system.

Superior intelligence in man is in a way related to size of the brain. Microcephalic individuals whose brains remain considerably under normal size are mentally defective, but an individual with an exceptionally large brain is not necessarily a genius. There are many causes that affect the size of the brain. The larger the body, the larger the brain. Therefore well-nourished people who have a larger bulk of body than those poorly nourished have larger brains, not because their brains are structurally more highly developed, but because the larger bulk is a characteristic feature of the entire bodily form. Eminent people belong generally to the better nourished class and the cause of the greater brain is, therefore, uncertain. The variation in the size of the brain of eminent men is also very considerable, some falling way beneath the norm.

The real problem to be solved is the relation between the structure of the brain and its function. The correlation between gross structure in the races of man and function is so slight that no safe inferences may be drawn on the basis of the slight differences between races which are of such character that up to this time the racial identification of a brain is impossible, except in so far as elongated and rounded heads, high and low heads and similar gross forms may be distinguished which do not seem to have any relation to minute structure or function. At least it has never been proved to exist and it does not seem likely that there is any kind of intimate relation.

The differences between races are so small that they lie within the narrow range in the limits of which all forms may function equally well. We cannot say that the ratio of inadequate brains and nervous systems, that function noticeably worse than the norm, is the same in every race, nor that those of rare excellence are equally frequent. It is not improbable that such differences may exist in the same way as we find different ranges of adjustability in other organs.

If the anatomical structure of the brain is a doubtful indication of mental excellence, this is still more the case with differences in other parts of the body. So far as we can judge, the form of the foot and the slight development of the calves of the Negro; the prominence of his teeth and the size of his lips; the heaviness of the face of the Mongol; or the difference in degree of pigmentation of the races have no relation to mentality. At least every attempt to prove such relation has failed.

In any attempt to place the human races in an evolutionary series we must also remember that modern races are not wild but domesticated forms. In regard to nutrition and artificial protection the mode of life of man is like that of domesticated animals. The artificial modification of food by the use of fire and the invention of tools were the steps that brought about the self-domestication of man. Both belong to a very early period, to a time before the last extensive glaciation of Europe. Man must be considered the oldest domesticated form. The most characteristic features of human races bear evidence of this. The loss of pigmentation in the blond, blue-eyed races; the blackness of the hair of the Negro are traits that do not occur in any wild mammal form. Exceptions are the blackness of the hair coat of the black panther, of the black bear and of the subterranean mole. The frizzliness of the Negro hair and the curliness of the hair of other races, the long hair of the head, do not occur in wild mammals. The permanence rather than periodicity of the sexual functions and of the female breast; the anomalies of sexual behavior are in most cases characteristics of domesticated animals. The kind of domestication of man is like that of the animals raised by primitive tribes that do not breed certain strains by selection. Nevertheless, forms differing from the wild forms develop in their herds.

Some of the traits of man that might be considered as indicating a lower evolutionary stage may as well be due to domestication. Reduction or unusual lengthening of the face occur. The excessive reduction of the face in some White types and the elongation of the mouth parts of the Negro may be due to this cause. It may be a secondary development from an intermediate form. The brain of domesticated forms is generally smaller than that of wild forms. In exceptional cases it may be larger. Pygmy forms and giants develop in domestication. The so-called "primitive traits" of races are not necessarily indications of an early arrest. They may be later acquisitions stabilized in domestication.

All this, however, has little to do with the biologically determined mentality of races, which is often assumed to be the basis of social behavior. Mental behavior is closely related to the physiological functioning of the body and the problem may be formulated as an investigation of the functioning of the body, in the widest sense of the term "functioning."

We have seen that the description of the anatomical traits of a race in general terms involves a faulty generalization based on the impression made by the majority of individuals. This is no less true in regard to the functions, and particularly the mental functions, of a population. Our characterization of the mentality of a people is merely a conceptualization of those traits that are found in a large number of individuals and that are, for this reason, impressive. In another population other traits impress themselves upon the mind and are conceptualized. This does not prove that, if in a third population both types are found, its functional behavior is due to a mixed origin. The objective value of generalizations of this type is not self-evident, because they are merely the result of the subjective construction of types, the wide variability of which is disregarded.

Actually the functions exhibited by a whole race can be defined as hereditary even less than its anatomical traits, because individually and in family lines the variations are so great that not all the members of the race react alike.

When the body has completed its growth its features remain the same for a considerable length of time,—until the changes due to old age set in. It does not matter at what time we examine the body, the results will always be nearly the same. Fluctuations of weight, of the amount of fat, of muscle do occur, but these are comparatively slight, and under normal conditions of health, nutrition and exercise, insignificant until senility sets in.

It is different with the functions of the body. The heart beat depends upon transient conditions. In sleep it is slow; in waking, during meals, during exercise more rapid. The range of the number of heart beats for the individual is very wide. The condition of our digestive tract depends upon the amount and kind of food present; our eyes act differently in intense light and in darkness. The variation in the functions of an individual is considerable. Furthermore, the individuals constituting a population do not all function in the same way. Variability, which in regard to anatomical traits has only one source, namely, the differences between individuals, has in physiological functions an added source, the different behavior of the individual at different times. It is, therefore, not surprising that functionally the individuals composing a population exhibit a considerable variability.

The average values expressing the functioning of various races living under the same conditions are not the same, but the differences are not great as compared to the variations that occur in each racial group. Investigations of the functioning of the same sense organs of various races, such as Whites, Indians, Filipinos and people of New Guinea, indicate that their sensitiveness is very much the same. The popular belief in an unusual keenness of eyesight or hearing of primitive people is not corroborated by careful observations. The impression is due to the training of their power of observation which is directed to phenomena with which we are not familiar. Differences that may be significant have been found in the basal metabolism of Mongols and Whites, but these are contradicted by observations made on natives of Yucatan. While East Asiatics, residents of the United States, showed on the average a low value of basal metabolism that of the Yucatec in Yucatan was high. There are probably differences between Whites and Negroes in the functioning of the digestive tract and of the skin. Much remains to be done in the study of physiological functions of different races before we can determine the quantitative differences between them.

The variability of many functions is well known. We referred before to the heart beat. Let us imagine an individual who lives in New York and leads a sedentary life without bodily exercise. Transport this person to the high plateaus of the Bolivian Andes where he has to do physical work. He will find difficulties for a while, but, if he is healthy, he will finally become adjusted to the new conditions. His normal heart beat, however, will have changed. His lungs also will act differently in the rarefied air. It is the same individual who in the new environment will exhibit a quantitatively different functioning of the body.

The condition is analogous to the one found in the variability of bodily form of lower organisms which is subject to important modifications brought about by the environment. The functions of the organs are adjustable to different requirements. Every organ has—to use Dr. Meltzer's term—a margin of safety. Within limits it can function normally according to environmental requirements. Even a partly disabled organ can be sufficient for the needs of the body. Inadequacy develops only when these limits are exceeded. There are certain conditions that are most favorable, but the loss of adequacy is very slight when the conditions change within the margins of safety.

In most cases of the kind here referred to the environmental influence acts upon different individuals in the same direction. If we bring two organically different individuals into the same environment they may, therefore, become alike in their functional responses and we may gain the impression of a functional likeness of distinct anatomical forms that is due to environment, not to their internal structure. Only in those cases in which the environment acts with different intensity or perhaps even in different directions upon the organism may we expect increased unlikeness under the same environmental conditions. When, for instance, for one individual the margin of safety is so narrow that the environmental conditions are excessive, for another one so wide that adequate adjustment is possible, the former will become sick, while the other will remain healthy. Davenport has called attention to a typical case of this kind, when two individuals of similar complexion are exposed to sunlight, the one may develop red color, the other may tan brown.

What is true of the physiological functioning of the body is still more true of mental reactions. A simple example may illustrate this. When we are asked to react to a stimulus, for instance by tapping in response to a signal given by a bell, we can establish a certain basal or minimum time interval between signal and tapping which is found when we are rested and concentrate our attention upon the signal. As soon as we are tired and when our attention is distracted the time increases. We may even become so much absorbed in other matters that the signal will go unnoticed. Environmental conditions determine the reaction time. The basal time for two individuals may differ quite considerably, still under varying environmental conditions they will react in the same way. If the conditions of life compel the one to concentrate his attention while the other has never been required to do so, they may react in the same way, although structurally they represent different types.

In more complex mental and social phenomena this adjustment of different types to a common standard is of frequent occurrence. The pronunciation of individuals in a small community is so uniform that an expert ear can identify the home of a person by his articulation. Anatomically the forms of the mouth, inner nose and larynx of all the individuals participating in this pronunciation vary considerably. The mouth may be large or small, the tongue thin or thick, the palate arched or flat. There are differences in the pitch of the voice and in timbre. Still the dialect will be the same for all. The articulation does not depend to any considerable extent upon the form of the mouth, but upon its use.

In all our everyday habits imitation of habits of the society to which we belong exerts its influence over the functioning of our minds and bodies and a degree of uniformity of thought and action is brought about among individuals who differ considerably in structure.

It would not be justifiable to claim that bodily form has no relation whatever to physiological or mental functioning. I do not believe that Watson is right when he claims that the mental activities of man are entirely due to his individual experiences and that what is called character or ability is due to outer conditions, not to organic structure. It seems to me that this goes counter to the observation of mental activities in the animal world as well as among men. The mental activities of a family of idiots will not, even under the most favorable conditions, equal those of a highly intelligent family, and what is true in this extreme case must be true also when the differences are less pronounced. Although it is never possible to eliminate environmental influences that bring about similarity or dissimilarity, it seems unreasonable to assume that in the mental domain organically determined sameness of all individuals should exist while in all other traits we do find differences; but we must admit that the organic differences are liable to be overlaid and overshadowed by environmental influences.

Under these conditions it is well-nigh impossible to determine with certainty the hereditary traits in mental behavior. In a well-integrated society we find people of most diverse descent who all react so much in the same way that it is impossible to tell from their reactions alone to what race they belong. Individual differences and those belonging to family lines occur in such a society, but among healthy individuals these are so slightly correlated with bodily form that an identification of an individual on the basis of his functions as belonging to a family or race of definite hereditary functional qualities is also impossible.

In this case, even more than in that of anatomical form, the range of variation of hereditary lines constituting a "race" is so wide that the same types of lines may be found in different races. While so far as anatomical form is concerned Negroes and Whites have hereditary racial traits, this is not true of function. The mental life of each of the individuals constituting these races is so varied that from its form alone an individual cannot be assigned to the one or the other. It is true that in regard to a few races, like the Bushmen of South Africa, we have no evidence in regard to this point, and we may suspend judgment, although I do not anticipate that any fundamental differences will be found.

So far as our experience goes we may safely say that the differences between family lines are much greater than the differences between races. It may happen that members of one family line, extreme in form and function, are quite different from those of a family line of the opposite extreme, although both belong to the same race; while it may be very difficult to find individuals or family lines in one racial type that may not be duplicated in a neighboring type.

The assumption of fundamental, hereditary mental characteristics of races is often based on an analogy with the mental traits of races of domesticated animals. Certainly the mentality of the poodle dog is quite different from that of the bulldog, or that of a race horse from that of a dray horse.

This analogy is not well founded, because the races of domesticated animals are comparable to family lines, not to human races. They are developed by carefully controlled inbreeding. Their family lines are uniform; in man they are diverse. The types constituting breeds are parallel to the family lines that occur in all human races, which, however, do not become stabilized on account of the lack of rigid inbreeding. In this respect human races must be compared to wild animals, not to selected, domesticated breeds.

All these considerations are apparently contradicted by the results of the so-called intelligence tests which were originally intended to determine innate intellectuality. Actually these tests show considerable differences not only between individuals but also between racial and social groups. The test is an expression of mental function. Like other functions the responses to mental tests show overlapping of individuals belonging to different groups and ordinarily it is not possible to assign an individual to his proper group according to his response.

The test itself shows only that a task set to a person can be performed by him more or less satisfactorily. That the result is solely or primarily a result of organically determined intelligence is an assumption that has to be proved. Defective individuals cannot perform certain acts required in the tests. Within narrower limits of performance we must ask in how far the structure of the organism, in how far outer, environmental conditions may determine the result of the test. Since all functions are strongly influenced by environment it is likely that here also environmental influences may prevail and obscure the structurally determined part of the reaction.

Let us illustrate this by an example. One of the simplest tests consists in the task of fitting blocks of various forms into holes of corresponding forms. There are primitive people who devote much time to decorative work in which fitting of forms plays an important part. It may be appliqué work, mosaic, or stencil work. Others have no experience whatever in the use of forms.

Dr. Klineberg has tested the ability of Indian girls who were still somewhat familiar with the old style of bead work, in regard to their ability to reproduce geometrical forms of varying complexity. He found that the girls among the Sac and Fox, a tribe in which bead work is still alive, had the greatest ability to reproduce forms. Next to them were the Dakota girls who were markedly superior to White girls. All girls were much superior to boys, Whites and Indians, who are not familiar with bead work. Experience enabled the girls to grasp new, previously unknown forms rapidly and easily.

He has also investigated the reactions to simple tests of various races living under very different conditions. He found that all races investigated by him respond under city conditions quickly and inaccurately, that the same races in remote country districts react slowly and more accurately. The hurry and pressure for efficiency of city life result in a different attitude that has nothing to do with innate intelligence, but is an effect of a cultural condition.

An experiment made in Germany, but based on entirely different sets of tests, has had a similar result. Children belonging to different types of schools were tested. The social groups attending elementary schools and higher schools of various types differ in their cultural attitudes. It is unlikely that they belong by descent to different racial groups. On the contrary, the population as a whole is fairly uniform. The responses in various schools were quite different. There is no particular reason why we should assume a difference in organic structure between the groups and it seems more likely that we are dealing with the effects of cultural differentiation.

In all tests based on language the effect of the linguistic experience of the subject plays an important part. The familiarity with a language, the ease of understanding what is demanded in the test has a decided influence upon the result. This may be accentuated when the test is given in a foreign or any imperfectly acquired language. Besides this, our whole sense experience is classified according to linguistic principles and our thought is deeply influenced by the classification of our experience. Often the scope of a concept expressed by a word determines the current of our thought and the categories which the grammatical form of the language compels us to express keep certain types of modality or connection before our minds. When language compels me to differentiate sharply between elder and younger brother, between father's brother and mother's brother, directions of thought that our vaguer terms permit will be excluded. When the terms for son and brother's son are not distinguished the flow of thought may run in currents unexpected to us who differentiate clearly between these terms. When a language states clearly in every case the forms of objects, as round, long or flat; or the instrumentality with which an action is done, as with the hand, with a knife, with a point; or the source of knowledge of a statement, as observed, known by evidence or by hearsay, these forms may establish lines of association. Comparison of reactions of individuals that speak fundamentally distinct languages may, therefore, express the influence of language upon the current of thought, not any innate difference in the form of thought.

All these considerations cause us to doubt whether it is possible to differentiate between environmental and organic determination of responses, as soon as the environment of two individuals is different.

It is exceedingly difficult to secure an identical environment even in our own culture. Every home, every street, every family group and school has its own character which is difficult to evaluate. In large masses of individuals we may assume a somewhat equal environmental setting for a group in similar economic and social position, and it is justifiable to assume in this case that the variability of environmental influence is much restricted and that organically determined differences between individuals appear more clearly.

Just as soon as we compare different social groups the relative uniformity of social background disappears and, if we are dealing with populations of the same descent, there is a strong probability that differences in the type of responses are primarily due to the effect of environment rather than to organic differences between the groups.

The responses to tests may be based on recognition of sensory impressions, on motor experience, such as the results of complex movements; or on the use of acquired knowledge. All of these contain experience. A city boy who has been brought up by reading, familiar with the conveniences of city life, accustomed to the rush of traffic and the watchfulness demanded on the streets has a general setting entirely different from that of a boy brought up on a lonely farm, who has had no contact with the machinery of modern city life. His sense experience, motor habits and the currents of his thoughts differ from those of the city boy.

Certainly in none of the tests that have ever been applied is individual experience eliminated and I doubt that it can be done.

We must remember how we acquire our manner of acting and thinking. From our earliest days we imitate the behavior of our environment and our behavior in later years is determined by what we learn as infants and children. The response to any stimulus depends upon these early habits. Individually it may be influenced by organic, hereditary conditions. In the large mass of a population these vary. In a homogeneous social group the experience gained in childhood is fairly uniform, so that its influence will be more marked than that of organic structure.

The dilemma of the investigator appears clearly in the results of mental tests taken on Negroes of Louisiana and Chicago. During the World War the enlisted men belonging to the two groups were tested and showed quite distinct responses. There is no very great difference in the pigmentation of the two groups. Both are largely mulattoes. The Northern Negroes passed the tests much more successfully than those from the South. Chicago Negroes are adjusted to city surroundings. They work with Whites and are accustomed to a certain degree of equality, owing to similarity of occupation and constant contact. All these are lacking among the Louisiana rural Negroes. Dr. Klineberg has shown what is actually happening. He studied the results of intelligence tests applied to Negroes who had moved from the country to the city and also to those who had moved from southern, more leisurely communities, to New York. He found that within a number of years they became adjusted to their new environment. While the results of the tests taken on those who had just moved to the city or to New York showed low averages, those who had lived in the cities or in New York showed the better results the longer they had lived in their new environment. The reason must be looked for in the character of the tests which are based on the experiences of city life and not on that of a rural community.

It has been claimed that the observed differences between rural and urban populations are due to selective migration, that a more energetic and intelligent group of Negroes has migrated to the cities and to New York and that the weak and unintelligent have stayed behind. Dr. Klineberg has tested this assumption in a number of cases and has compared the results of intelligence tests of those who stayed behind and of the migrants, taken before their migration. The results do not show any appreciable difference between the two groups, rather a very slight, presumably insignificant advantage for those who did not migrate.

It seems gratuitous to disregard the effect of social environment. We know that the environment is distinct and that human behavior is strikingly modified by it. According to the few tests made selection plays no important part in the migration of the Southern Negro to Northern cities. It is quite arbitrary to ascribe the difference in mental behavior solely to the latter, doubtful cause and to disregard the former entirely. Those who claim that there is an organic difference must prove it by showing the differences between the two groups before their migration.

Even if it were true that selection accounts for the differences in the responses to tests among these two groups, it would not have any bearing upon the problem of racial characteristics, for we should have here merely a selection of better endowed individuals or family lines, all belonging to the same race, a condition similar to the often quoted, but never proved, result of the emigration from New England to the West. The question would still remain, whether there is any difference in racial composition in the two groups. So far as we know the amount of Negro and White blood in the two groups is about the same.

Other tests intended to investigate differences between the mental reactions of Negroes, Mulattoes and Whites due to the racial composition of the groups are not convincing, because due caution has not been taken to insure an equal social background. The study of mental achievement of a socially uniform group undertaken by Dr. Herskovits does not show any relation between the intensity of negroid features and mental attainment. Up to this time none of the mental tests gives us any insight into significant racial differences that might not be adequately explained by the effect of social experience. Even Dr. Woodworth's observations on the Filipino pygmies are not convincing, because the cultural background of the groups tested is unknown.

A critical examination of all studies of this type in which differences between racial groups in regard to mental reactions are demonstrated, leaves us in doubt whether the determining factor is cultural experience or racial descent. We must emphasize again that differences between selected groups of the same descent, such as between poor orphan children, often of defective parentage, and of normal children; and those between unselected groups of individuals representing various races are phenomena quite distinct in character. In the former case the results of tests may express differences in family lines. Similar peculiarities might be found, although with much greater difficulty, when comparing small inbred communities, for inbred communities are liable to differ in social behavior. For large racial groups acceptable proof of marked mental differences due to organic, not social, causes has never been given.

Students of ethnology have always been so much impressed by the general similarity of fundamental traits of human culture that they have never found it necessary to take into account the racial descent of a people when discussing its culture. This is true of all schools of modern ethnology. Edward B. Tylor and

Herbert Spencer in their studies of the evolution of culture, Adolf Bastian in his insistence on the sameness of the fundamental forms of thought among all races, Lewis Morgan in his study of social forms, Westermarck in his inquiries into the history of moral ideas and of marriage—they all have carried on their work without any regard to race.

Friedrich Ratzel, who followed the historical dissemination of cultural forms does not pay attention to race, except in so far as he sometimes falls back upon vague mental characteristics of racial groups, a belief which he inherited from the older school of deductive ethnologists like Klemm and Carus. It may also be recognized that those investigators who try to reconstruct exceedingly ancient primitive cultural strata, like Graebner, Pater Schmidt and Dr. Koppers, are easily led to associate these with fundamental racial groups, without, however, giving any proof of the way in which social traits are dependent upon racial character.

The general experience of ethnologists who deal with recent ethnological phenomena indicate that whatever organic differences between the great races there may be, they are insignificant when considered in their effect upon cultural life.

It does not matter from which point of view we consider culture, its forms are not dependent upon race. In economic life and in regard to the extent of their inventions the Eskimos, the Bushmen and the Australians may well be compared. The position of the Magdalenian race, which lived at the end of the ice age, is quite similar to that of the Eskimo. On the other hand, the complexities of inventions and of economic life of the Negroes of the Sudan, of the ancient Pueblos, of our early European ancestors who used stone tools, and of the early Chinese are comparable.

In the study of material culture we are constantly compelled to compare similar inventions used by people of the most diverse descent. Devices for throwing spears from Australia and America; armor from the Pacific Islands and America; games of Africa and Asia; blowguns of Malaysia and South America; decorative designs from almost every continent; musical instruments from Asia, the Pacific Islands and America; head rests from Africa and Melanesia; the beginning of the art of writing in America and in the Old World; the use of the zero in America, Asia and Europe; the use of bronze, of methods of firemaking in many parts of the world cannot be studied on the basis of their distribution by races, but only by their geographical and historical distribution, or as independent achievements, without any reference to the bodily forms of the races using these inventions.

Other aspects of cultural life are perhaps still more impressive, because they characterize the general cultural life more deeply than inventions: the use of standards of value in Africa, America, Asia, Europe and on the islands of the Pacific Ocean; analogous types of family organization, such as small families, or extended sibs with maternal or paternal succession; totemic ideas; avoidance of close relatives; the exclusion of women from sacred ceremonials; the formation of age societies; all these are found in fundamentally similar forms among all races. In their study we are compelled to disregard the racial position of the people we study, for similarities and dissimilarities have no relation whatever to racial types.

It does not matter how the similar traits in diverse races may have originated, by diffusion or independent origin. They convince us of the independence of race and culture because their distribution does not follow racial lines.

CHAPTER III

THE INTERRELATION OF RACES

WE HAVE seen that from a purely biological point of view the concept of race unity breaks down. The multitude of genealogical lines, the diversity of individual and family types contained in each race is so great that no race can be considered as a unit. Furthermore, similarities between neighboring races and, in regard to function, even between distant races are so great that individuals cannot be assigned with certainty to one group or another.

Nevertheless, race consciousness exists and we have to investigate its source. It is customary to speak of an instinctive race consciousness. Even Romain Rolland says of it, "Ce vieux levain d'antipathie instinctive, qui couve au fond des cœurs de tous les hommes du Nord pour les hommes du Midi."

The feeling between Whites and Negroes in our country is decidedly of this character. There is an immediate feeling of contrast that is expressed in the popular conviction of the superiority of the White race. The feeling extends even to cases in which the Negro admixture is very slight and in which there is no certainty of the racial position of the individual. Proof of this are the numerous divorce suits based on alleged Negro descent. In this case the popular belief in the possible reversion of the offspring to a pure Negro type may be a determinant. This consideration does not enter law suits instituted to set aside adoption of children on account of their racial descent; or in the difficulties experienced by child-placing agencies which endeavor to find homes for children of suspected Negro descent,—no matter how little this may be expressed in their outer appearance.

It is necessary to make clear to ourselves what we mean when we speak of instinctive race consciousness.

We have to inquire whether race consciousness and race antipathies are truly instinctive or whether they are established by habits developed in childhood.

The basis of race consciousness and race antipathies is the dogmatic belief in the existence of well-defined races all the members of which possess the same fundamental bodily and mental characters. The results which we have reached in regard to the lack of clarity of the concept of race induces us to inquire whether these feelings are universal and whether other types of groups develop analogous feelings of contrast.

Race consciousness differs considerably in intensity. In the United States, taken as a whole, the feeling of aloofness between White and Negro is strongest. On the Pacific coast it is locally equalled by the feeling of the Whites against Asiatics and Indians. The feeling against the Japanese is most strikingly manifested by the enactment of a law forbidding marriages between Whites and Japanese. It has led to the anomalous position of American-born children of Japanese parents who have become completely Americanized and who nevertheless have no place in the White Community.

I have been told by those familiar with conditions in Humboldt County, California, that the White settlers will readily eat with Negroes, but not with Indians. In general, feeling of aversion to the Indian is rather slight. There is even a marked tendency of individuals with admixture of Indian blood to be proud of their ancestry, at least until recently, when early intermingling of Negroes and eastern Indians became better known.

Race feeling between Whites, Negroes, and Indians in Brazil seems to be quite different from what it is among ourselves. On the coast there is a large Negro population. The admixture of Indian is also quite marked. The discrimination between these three races is very much less than it is among ourselves, and the social obstacles for race mixture or for social advancement are not marked. Similar conditions prevail on the island of Santo Domingo where Spaniards and Negroes have intermarried. Perhaps it would be too much to claim that in these cases race consciousness is nonexistent; it is certainly much less pronounced than among ourselves.

If it is true that race antipathy among different

groups of mankind takes distinctive forms and expresses itself with varying intensity, we may doubt whether we are dealing with an instinctive phenomenon.

It will be found advantageous to investigate similar phenomena in the animal world. We know the peculiar antipathies between certain animals, such as dog and cat, horse and camel. These are organically determined, although they may be individually overcome. They might be considered analogous to the feeling between races if we had the same instinctive hostility or fear between individuals of different human races; but this has never been observed. On the contrary, under favorable conditions the reaction seems to be one of friendly curiosity.

The first view of an entirely foreign type is likely to impress us with consciousness of contrast, that may well take the form of antipathy. An example of this is the terror which the blue-eyed blond hordes of Ariovistus spread among the Roman legions. The first reaction to strange appearance should not be mistaken for race antipathy for it is strictly analogous to the revulsion against ugliness of appearance, strong body odor, deformities or even bad manners occurring in our midst. They are not determined by race but by certain esthetic standards that determine our preferences and antipathies. Constant familiarity with strange types modifies our standards to such an extent that the consciousness of contrast becomes very slight. The examples given before illustrate this process.

Conditions analogous to those found in racial groups occur in animal societies. Gregarious animals live either in open or in closed societies. Open societies are those in which any outside individual may join a herd. They are found among mammals and birds, but particularly among fishes, insects, and other lower animals. A swarm of mosquitoes, a shoal of fish keep together but do not exclude newcomers of the same species, sometimes even of other species. Herds of ruminants are often organized under leaders but may not exclude newcomers. The behavior of animals that occupy a definite area as their feeding ground is quite different. They treat every newcomer as an enemy and while he may succeed in gaining admission after a number of combats, the first endeavor of the herd is to drive away or to kill the intruder. Many herds of monkeys are said to behave in this way. Penguins on their breeding places will drive away stray visitors, while admitting their neighbors. The best known example is that of the Pariah dogs of Oriental towns. The dogs of one street did not admit one from another street and the stranger was killed by them if he did not beat a hasty retreat. The most perfect forms of closed societies are found in the insect states. Ants of the same hill recognize one another by the scent of the hill and attack every strange ant. Even insects of another species, if only they participate in the scent of the particular hill, are welcomed. Sameness of species does not decide the attitude towards the individual. Participation in the scent of the hill is the feature by which membership in the group is determined.

The groups do not need to be related by descent. They may be thrown together by accident. Nevertheless, according to the habits of the species, they will form a closed society.

In primitive human society every tribe forms a closed society. It behaves like the Oriental Pariah dogs.

In the early days of mankind our earth was thinly settled. Small groups of human beings were scattered here and there; the members of each horde were one in speech, one in customs, one in superstitious beliefs. In their habitat they roamed from place to place, following the game that furnished their subsistence, or digging roots and picking the fruits of trees and bushes to allay the pangs of hunger. They were held together by the strong bands of habit. The gain of one member of the horde was the gain of the whole group, the loss and harm done to one was loss and harm to the whole community. No one had fundamental interests at stake that were not more or less also the interests of his fellows.

Beyond the limits of the hunting grounds lived other groups, different in speech, different in customs, perhaps even different in appearance, whose very existence was a source of danger. They preyed upon the game, they threatened inroads upon the harvest of roots and fruits. They acted in a different manner; their reasoning and feeling were unintelligible; they had no part in the interests of the horde. Thus they stood opposed to it as beings of another kind, with whom there could be no community of interest. To harm them, if possible to annihilate them, was a self-evident act of self-preservation.

Thus the most primitive form of society presents to us the picture of continuous strife. The hand of each member of one horde was raised against each member of all other hordes. Always on the alert to protect himself and his kindred, man considered it an act of high merit to kill the stranger.

The tendency to form closed societies is not by any means confined to primitive tribes. It exists to a marked extent in our own civilization. Until quite recent times, and in many cases even now, the old nobility formed a closed society. The patricians and plebeians, Greeks and barbarians, the gangs of our streets, Mohammedans and infidels,—and our own modern nations are in this sense closed societies that cannot exist without antagonisms.

The principles that hold societies together vary enormously, but common to all of them is the feeling of antagonism against other parallel groups.

Racial groups differ in one respect from the societies here enumerated. While the position of an individual as a member of one of the socially determined groups is not evident, it is apparent when the grouping is made according to bodily appearance. If the belief should prevail, as it once did, that all red-haired individuals have an undesirable character, they would at once be socially segregated and no red-haired person could escape from his class. The Negro who may at once be recognized by his bodily build is automatically placed in his class and not one of them can escape from the effect of being excluded from the closed group of Whites.

When individuals are to be herded together in a closed group the dominant group may prescribe for them a distinguishing symbol,—like the garb of the medieval Jews or the stripes of the convict,—so that each individual who may otherwise have no distinguishing characteristic, may at once be assigned to his group and treated accordingly.

The assignment to a closed group may also be effected by a classifying name, like the term Dago for Italians which is intended to evoke the thought of all the supposed characteristics that are without reflection ascribed to all the members of the nation. Perhaps one of the most striking illustrations of this tendency in the present life of the United States is the assignment of anyone with a Jewish name to an undesirable group whose members are, according to the fancy of the owner, not allowed to dwell in certain buildings, not admitted in hotels or clubs and are in other ways discriminated against by the unthinking, who can see in the individual solely the representative of a class.

We have seen that from a biological point of view there is no reason for drawing a clean-cut line between races, because the lines of descent in each are physiologically and psychologically diverse, and because functionally similar lines occur in all races.

The formation of the racial groups in our midst must be understood on a social basis. In a community comprising two distinct types which are socially clearly separated, the social grouping is reënforced by the outer appearance of the individuals and each is at once and automatically assigned to his own group. In other communities,—as among Mohammedans or in Brazil,—where the social and racial groupings do not coincide, the result is different. The socially coherent groups are racially not uniform. Hence the assignment of an individual to a racial group does not develop as easily, the less so the more equal the groups in their social composition. A characteristic case occurs in South Africa where Whites, Negroes and Malay from southern Asia form three distinct groups. Suaheli Negroes who are Mohammedans like the Malay do not belong to the Negro group but to the Malay.

Dr. Manuel Andrade, in a personal communication, describes the interracial conditions in the Dominican Republic as follows: "There is no restaurant, hotel, or club in which color distinctions are observed. Government positions, of course, are open to all, and we do find Negroes and Mulattoes in all classes of government posts, including the presidency.

"I had occasion to see a review of a portion of the army. The main officer was almost White, but most of the captains and lieutenants were Negroes. On the other hand, there were several Whites among the common soldiers.

"In the evenings people promenade in the Plazas as in all Spanish-speaking countries. Married and engaged couples may be seen showing all varieties of color combination, including Negroes with light complexioned women. My impression is, however, that it is more frequent to find Black men married to White or nearly White women, than White men married to Black women. Economical considerations may be a factor here. A White woman may accept a dark man because of his lucrative occupation or political position.

"I was invited to a ball given by a social and literary club in the town of La Vega. The members and guests present showed the same range and variety of colors prevalent anywhere in the Republic. They were supposed to represent the best social elements in the town. Among the dancing couples, there were several extreme combinations of apparently pure Negro men and White women.

"The two ladies who own the Hotel de las Dos Hermanas in the city of Santiago de los Caballeros and their brother would be considered White anywhere. Especially the brother, who has light gray eyes and reddish hair. In the course of conversation he asked me in what part of Spain I was born. He proceeded to tell me from what part of Spain his father and his mother's father had come, and added humorously that were it not for the little African blood he had in his veins, he could very well say that he was my fellow-countryman. I think this candid reference to his African ancestors, in the unconcerned manner in which he said it, is a significant index to the prevalent feeling. We must consider he was trying to promote cordial relations toward a guest in his sisters' hotel. I find in this instance a corroboration of my general impression, that it makes no difference whether one has Negro blood in him, though it may make some difference for such admixture to be in evidence in his features or color.

"It seems to me that racial differences are felt to a certain extent in matters pertaining to marriage or sexual relations, but that the division is not one of pure White against other admixtures. The difference may be felt in proportion to the divergence in color, but the prevalence of marriages between Negroes and White women would seem to indicate that the objection is not very strong.

"In the current social intercourse between man and man I was not able to detect the slightest indication of prejudice. In one instance, a nearly White man trying to describe a certain individual had forgotten whether he was White or 'pardito'."

It is a characteristic feature of closed groups that the feeling of solidarity is expressed by an idealization of the group and by the desire for its perpetuation. When the groups are denominational, there is strong antagonism against marriages outside of the groups. The group must be kept pure, although denomination and descent are in no way related. If the social groups are racial groups we encounter in the same way the desire for racial exogamy in order to maintain racial purity. This, however, has no relation to sexual antipathy, for it is solely a result of social pressure. The weakening of race consciousness in communities in which children grow up as an almost homogeneous group; the occurrence of equally strong antipathies between denominational groups, or between social strata—as witnessed even now by the exclusiveness of European nobility and the Indian castes, in earlier times by the Roman patricians and plebeians, the Spartan Lacedemonians, Periœci and Helots, and the Egyptian castes—all these show that antipathies are social phenomena. The variety of incest groups which occur in human society also shows that sexual aversion is not organically determined but due to social customs. Otherwise it would be unintelligible why in some societies cousin marriages are shunned, in others prescribed, why among some tribes the young men and women of the same social group are forbidden, among others compelled to intermarry; why sometimes everybody is required to marry in his own generation, while in other cases no attention is paid to generation.

In all these cases there is no instinctive sexual aversion. Neither does it exist in the relation between Whites and Negroes. The free intermingling of slave owners with their female slaves and the resulting striking decrease in the number of full-blood Negroes is ample proof of the absence of any sexual antipathy. The rarity of the reverse intermixture, that of male Negroes and female Whites, can be fully understood on the basis of social conditions. In view of the behavior of the male White and of the forms of mixture in other societies it does not seem likely that it is reducible to innate sexual antipathy. The White master sought his colored mates who had little power to resist him. The colored slave was in an entirely different position towards his mistress and to other White women.

The intermingling of Indian and White throws an interesting light upon this subject. Owing to other reasons the early intermingling between the two races was also between White males and Indian females. It was caused not by the relation of master and slave woman but by the absence of White women. The general development has been such that Mestizo women—that is, those of Indian-White descent—are liable to marry Whites. Their descendants gradually pass out of the Indian population unless economic privileges, such as the right to hold valuable lands belonging to the Indians, serve as an attraction to the Indian community. The men, on the other hand, are more liable to marry Indian or Mestizo women and remain in the tribe. The male descendants of Mestizo women who no longer belong to a segregated group marry freely among the Whites, while the male descendants of Mestizo men are ordinarily not in the position to marry outside of their own race.

There is no doubt that the strangeness of a foreign racial type plays an important rôle in these relations. The ideal of beauty of a person who is growing up in an exclusively White society is different from that of a Negro who lives in a Negro society and the later in life a White person comes into contact with a Negro the more keenly will he be conscious of the strangeness of the type and, while there is curiosity, there is also reluctance to close association. The same attitude develops when racial and social groupings coincide, so that reluctance to entering into social contact may be reinterpreted as racial dislike.

Here again the question arises whether these influences would act in the same way if the groups were socially not separated. We can find an answer to this question solely by a consideration of conditions in countries in which there is no pronounced race feeling. It would seem that there the attractiveness of forms has a much wider range, and is not determined by pigmentation and other racial traits alone. Aversion is not expressed on racial lines but on the ground of the repulsiveness of other features. Preferences and aversions differ individually.

Unfortunately these conditions cannot be proven by actual numerical observations that would be convincing. All we can give are the results of general observations. These are, however, so striking that their validity seems well established.

Since the abolition of slavery the intermingling of Negroes and Whites has taken a curious course. Legitimate and illegitimate mating between Whites and Negroes has undoubtedly decreased and we find essentially marriages among Negroes and Mulattoes. Dr. Melville J. Herskovits has collected statistics on this subject. He found that, on the average, dark individuals will marry those of dark, though slightly lighter complexion, light ones those of light, though slightly darker complexion. This indicates that there is a decided preference in the mating of those of similar color,—an expression of the transfer of our own race feeling to the colored people who live among us and participate in our culture. But, furthermore, the darker man marries on the average a lighter woman. Since there is no difference in the pigmentation of the two sexes this indicates a preference on the part of the men,—another manifestation of the adoption of our valuations by the Negroes.

The effect of this selective process, if it continues for many generations, will be the passing of many of the lightest men out of the Negro community. Either they die as bachelors or they are merged in the general population. For the remainder it must inevitably lead to a darkening of the whole colored population, for the daughters of each generation, whose fathers are dark and whose mothers are light, will be darker than their mothers. When they again become mothers, their children will be still darker, provided the same conditions continue. Thus there will come to be a constantly increasing intensity of Negro characteristics and a sharper contrast between the two principal races of the country.

During the time of slavery the condition was the reverse. On account of the numerous unions between White men and Negro women the new generation was lighter than their mothers. A constant lightening of the Negro population resulted and hence a lessening of the racial contrast without any modification of the descendants of White females.

An evenly mixed population can result only if the number of matings between males of one race and females of the other is equal to that of matings in the opposite direction. Otherwise the racial type of the group descended in the female line will be unstable.

When social divisions follow racial lines, as they do among ourselves, the degree of difference between racial forms is an important element in establishing racial groupings and in accentuating racial conflicts. From this point of view the present tendency is most undesirable.

Under prevailing circumstances complete freedom of matrimonial union between the two races cannot be expected. The causes that operate against the unions of colored men and White women are almost as potent as in the days of slavery. Looking forward towards a lessening of the intensity of race feeling an increase of unions of White men and colored women would be desirable. The present policy of many of the Southern States tends to accentuate the lack of homogeneity of our nation.

The biological arguments that have been brought forward against race crossing are not convincing. Equally good reasons can be given in favor of crossings of the best elements of various races, and for closely related groups these arguments seem incontrovertible.

If we were to select the most intelligent, imaginative, energetic and emotionally stable third of mankind, all races would be represented. The mere fact that a person is a healthy European, or a blond European would not be proof that he would belong to this élite. Nobody has ever given proof that the mixed descendants of such a select group would be inferior.

If a selection of immigrants is to be made it should never be made by a rough racial classification, but by a careful examination of the individual and of his family history.

No matter how weak the case of racial purity may be, we cannot hope easily to overcome its appeal. As long as the social groups are racial groups we shall also encounter the desire for racial purity. When considerable racial differences are encountered in the same social group, they are disregarded unless there are introduced artificial ideals of bodily form that tend to establish new social divisions. This is occurring in some social groups in Europe and America who idealize the blond, blue-eyed type.

It follows that the "instinctive" race antipathy

can be broken down, if we succeed in creating among young children social groups that are not divided according to the principles of race and which have principles of cohesion that weld the group into a whole. Under the pressure of present popular feeling it will not be easy to establish such groups. Nevertheless, cultural coöperation cannot be reached without it.

Those who fear miscegenation, which I, personally, do not consider as in any way dangerous—not for the White race or for the Negro, or for mankind—may console themselves with their belief in a race consciousness, which would manifest itself in selective mating. Then matters would remain as they are.

CHAPTER IV

NATIONALISM

THE TERM "nationality" has two meanings. It is applied to designate collectively the citizens of one State, as when we describe a person's nationality as American, French, or Italian, meaning by this that he is a citizen of the United States, France, or Italy. It is also used to designate persons who belong to one linguistic and cultural group, as when we say that the many irregularly distributed communities of the Balkan Peninsula are of Bulgarian, Servian, Greek, or Turkish nationality.

The term "nation" is somewhat less ambiguous, for it is generally used to designate a political unit, a State, although it is also occasionally used collectively for the members of a nationality regardless of their political affiliations. Italians and Germans before the political unification of their countries were sometimes designated as the Italian or German nation.

The term "nationalism" is as ambiguous as the term "nationality." It is used to express the feeling of solidarity and of devotion to the interests of the State on the part of its citizens. It is also used to designate the desire of a nationality that feels its cultural unity for unity in political and economic organization.

In the following I use the term "nationality" to designate groups the same in culture and speech without reference to political affiliation. In this sense there are States that embrace several nationalities, like Czecho-Slovakia and Poland. A nationality may also be divided and constitute several States, like the Spaniards in a few States of South America, or the Italians and Germans before the unification of Italy and Germany; or the members of one nationality may be included in several States, like the Germans in Germany, Austria, France, Poland, Czecho-Slovakia, Italy and the Baltic States.

While the significance of the term "nationalism" is quite clear in so far as it relates to devotion to the interests of the State, it is not so clear in so far as it refers to the desires of a nationality, because there is little clarity in regard to the concept of nationality as a group characterized by unity of language and culture.

Since the general conditions of life prevailing in a State, particularly its institutions, mould to a certain extent the behavior of its citizens, the characteristics of a nationality are in part coincident with those of nations.

Furthermore the theory has been advanced that the cultural life of a people is dependent upon bodily build, and on this basis confusion between the concepts of race and of nationality as a linguistic and cultural group has arisen. In the terminology of the United States Immigration Commission English, French, German, and Russian are designated as races. In common parlance also no clear distinction is made between cultural groups and racial strains. The blond is supposed to represent the Teuton; the short and dark, Spaniard or Italian; the heavy built brunette, the Slav, and the observed characteristics of these groups are ascribed to their bodily build.

We have seen before (pp. 43 et seq.) that the existence of hereditary mental characteristics in large groups of man, particularly in closely allied groups, has never been satisfactorily established. Nevertheless the belief persists that a particular type and a correlated mentality are the characteristic elements among the great variety of forms that constitute a population which has in common cultural and linguistic traits. Thus it happens that the blond, blue-eyed type is considered as endowed with energy, intelligence and other traits that make him the real bearer of the culture of northwestern Europe and the true representative of northwest European nationalities.

It has been claimed that all the achievements of Greece are due to the blond immigrants who reached that country before the beginning of the historic era, although the presence of a blond element does not prove that its cultural advance was due to it. It might be said with equal justice that the rise of North European civilization did not begin until South and Central European blood became intermingled with that of the North European.

The same thought was in Haupt's mind when he tried to prove that Christ could not have been a Jew, but must have been by descent an Aryan, that means a North European; or Henry Fairfield Osborn who maintained that Columbus must have been a blond, or Sir Henry Keith who contrasted the types of Lord Kitchener and Hindenburg and assigned the difference in type as the cause of their supposed mental qualities.

This erroneous identification of a race as the true representative of a culture within a nationality, the assumption of a close correlation between race and culture has taken hold of the mind wherever the Teutonic, German, or Anglo-Saxon type—however it may be called—prevails; or where the Italian "race" glories in its past greatness and virtues.

Although Europeans begin to understand that each nationality embraces individuals of many different types, the belief prevails that in this mixture certain pure types continue to persist which possess qualities that make them the true bearers of national culture. Local "races" among which these "pure" types have disappeared or are disappearing are believed to be in danger of losing their national culture and the ideal type is admonished to see to it that it may not be swamped by so-called inferior types and that it preserve its purity and with it its national culture. Examples of this are the associations in Germany that admit only blond members, and the more numerous ones that exclude Jews.

The notion prevails among ourselves with equal force, for we are haunted by fear of the ominous influx of "inferior" races from eastern and southern Europe, of the mongrelization of the American people by intermixture with these types, because it is believed that we may lose in this way the characteristic mental traits that belong to the Northwest Europeans.

We should remember that people of pure descent or of a pure racial type are not found in any part of Europe. This is proved by the distribution of bodily forms. Even if it is true that the blond type is found at present preëminently among Teutonic people, it is not confined to them alone. Among the Finns, Poles, French, North Italians, not to speak of the North African Berbers and the Kurds of western Asia, there are individuals of this type. The heavy-set, darker East European type is common to many of the Slavic peoples of eastern Europe, to the Germans of Austria and southern Germany, to the North Italians, and to the French of the Alps and of central France. The Mediterranean type is spread widely over Spain, Italy, Greece, and the coast of Asia Minor, without regard to national boundaries. Other local types may be readily distinguished, if we take into consideration other differences in form. These are also confined to definite territories.

In western Europe, types are on the whole distributed in strata that follow one another from north to south,—in the north the blond, in the center a darker, short-headed type, in the south the slightly built Mediterranean.

National boundaries in central Europe, on the other hand, run north and south: and so we find many individuals in northern France, Belgium, Holland, Germany and northwestern Russia similar in type and descent; many of the central French, South Germans, Swiss, North Italians, Austrians, Servians and central Russians, belonging to similar varieties of man; and also persons in southern France, closely related to the types of the eastern and western Mediterranean area.

Ample historical evidence is available to show how this has come about. The relation of German and Slav is instructive. During the period of Teutonic migrations, in the first few centuries of our era, the Slavs settled in the region from which Teutonic tribes had moved away. They occupied the whole of what is now eastern Germany, but the population seems to have been sparse. In the Middle Ages, with the growth of the German Empire, a slow backward movement set in. Germans settled as colonists in Slavic territory, and by degrees German speech prevailed over the Slavic and a population of mixed descent developed. In Germany survivals of the gradual process may be found in a remote locality where Slavic speech still persists.

As by contact with the more advanced Germans the cultural and economic conditions of the Slavs improved and their numbers and their wealth increased, their resistance to Germanization became greater and greater,—earliest among the Czechs and Poles, later in the other Slavic groups. Later on, through a similar process, a mixed population of Poles, Lithuanians and Russians originated farther to the east.

This process has led to the present distribution of languages, which expresses a fossilization of German colonization in the east, and illustrates in a most striking way the penetration of peoples. Poland and part of Russia, Slavonic and Magyar territories are interspersed with small German settlements, which are the more sparse and scattered the farther east they are located, the more continuous the nearer they lie to Germany,—at least until the recent systematic persecution of Germans in Poland.

With the increased economic and cultural strength of the Slav, the German lost his ability to impose his mode of life upon him, and with it his power to assimilate the numerically stronger people in its own home. But by blood all these people, no matter what their speech, are the same.

A process analogous to the medieval Germanization of Slavic tribes may at present be observed in Mexico, where Indian speech and culture give way to Spanish. Each town forms a center of Spanish speech which, owing to the economic and cultural strength of the town, spreads over the surrounding country.

The French Huguenots who escaped from religious

persecution and settled in Germany have been completely assimilated, although the French school in which their children were educated is still in existence as a French gymnasium. Alsatians who migrated to Paris have become French in language and spirit; Germans have been absorbed by Russians; the Swedish nobility count among their numbers many descendants of the nobility of foreign countries. An analysis of the descent of the population of every part of Europe proves that intermingling has been going on for long periods.

The movements of tribes in prehistoric times and during antiquity also illustrate the ways in which different strains became mixed: the Doric migration into Greece, the movements of the Kelts into Spain, Italy and eastward as far as Asia Minor; the Teutonic migrations which swept through Europe from the Black Sea into Italy, France, Spain and on into Africa; the invasion of the Balkan Peninsula by Slavs, and their extension over eastern Russia and into Siberia; Phœnician, Greek and Roman colonization; the roving Normans; the expansion of the Arabs; the Crusades, are a few of the important events that have contributed to the intermingling of the European population.

In every single nationality of Europe the various elements of the continental population are represented. Proof that a selected type within a nationality is the carrier of definite mental and cultural traits has never been given. On the contrary, we find individuals of the same type but members of different nationalities behaving according to the national pattern, and individuals of the most diverse types, but members of the same nationality behaving in similar ways.

The readiness with which we recognize individuals, according to their outer appearance, as members of certain nationalities confirms this view. Such identifications, which are far from certain, are based only in part on the essential elements of the form of the body, such as hair and eye color, face form and stature. We are led much more by the mannerisms of wearing hair and beard, and by the characteristic expressions and motions of the body, which are determined not so much by hereditary causes as by habit. The latter are more impressive than the former; and among the nations of Europe no fundamental traits of the body occur that belong to one to the exclusion of the others. It is a common experience that Americans of European descent, French, Italian or German, are recognized as Americans, notwithstanding their pure descent and solely on account of their appearance and habits. These are expressions of their nationality, of their cultural life.

Racial descent has significance in determining nationality in those countries in which fundamentally distinct races live side by side. Everybody will agree that American Whites, Negroes and native-born Asiatics are members of the same nation, but they would hardly be called members of the same nationality, because of the social barriers between these groups and the consciousness that they are derived from races that continue to be distinct. They are separated by divergence in bodily form which causes, at least for the time being, permanent segregation. In Mexico, where the intermingling of Indian and White has produced a numerous mixed population which is not permanently separated by social barriers, the distinction between Indian, Mestizo and Spanish creole is weak and all are not only members of the Mexican nation, but also of Mexican nationality, provided they participate in the general social and political life of the country.

The social, not racial, significance of the term "nationality" appears also clearly in the position of the Jew in modern society. When the Jew is separated from the rest of the people among whom he lives by endogamy within the Jewish community, by habits, occupation and appearance, he is not entirely a member of the nationality, although a member of the nation, for he participates in part only in the interests of the community and endogamy keeps him permanently separated. When he is completely assimilated he is a member of the nationality. This appears most clearly in those North European countries in which the number of Jews is small and intermarriage and assimilation correspondingly rapid.

If community of racial descent is not the basis of nationality, is it community of language?

When we glance at the national aspirations that have characterized a large part of the nineteenth century, community of language might seem to be the background of national life. It touches the most sympathetic chords in our hearts. Italians worked for the overthrow of the small local and great foreign interests that were opposed to the national unity of all Italian-speaking people. German patriots strove and will strive for the federation of the German-speaking people in one empire. The struggles in the Balkans are largely due to a desire for national independence according to the limits of speech. The Poles have for more than a century longed for a reëstablishment of their state which is to embrace all those of Polish tongue.

It is, however, not very long that the bonds of language have been felt so intensely. Language establishes a basis of mutual understanding on which a community of interests may arise. The pleasure of hearing one's own tongue spoken in a foreign country creates at once between its speakers a feeling of comradeship that is quite real, and strong in proportion to the smallness of the number of speakers of the idiom. The necessity of easy communication between the members of one nation has also led generally to the endeavor to make one language the ruling language throughout the whole state. When there is a great difference of languages, as in the former Austria-Hungary, the national unity is liable to be feeble.

Notwithstanding unity of language severe internal conflicts may arise that do not allow the feeling for the unity of a nationality to arise. It may be all but lost owing to local or social conflicts, as in the case of the ancient Greek and the medieval cities; in differences between religious and cultural tradition, as among the Croatians and Servians; in social revolution; or in wars of religion.

Unity of language is more an ideal than a real bond; not only that divergence of dialects makes communication difficult, but community of thought among the members of different social classes is also so slight that no communication of deeper thought and feeling is possible. The Provençal and the North French, the Bavarian and the Westphalian peasant, the Sicilian and the Florentine are hopelessly divided, owing to differences of language. The transition of Italian into French is so gradual that only the political boundaries and the language imposed by Government, school and cultural relations determines whether we count a district as Italian or French. Unity is found in the educated groups that share the same language and the same emotional reactions.

In many ways the educated Americans, Englishmen, Frenchmen, Germans, Italians, Spanish, and Russians have more in common than each has with the uneducated classes of his own nation.

Neither the bonds of blood nor those of language alone make a nationality. It is rather the community of emotional life that rises from our everyday habits, from the forms of thoughts, feelings, and actions, which constitute the medium in which every individual can unfold freely his activities.

Language and nationality are so often identified, because we feel that among a people that use the same language every one can find the widest field for unrestricted activity. Thus is created the feeling for the existence of a national unit. Nevertheless it is perfectly clear that there is no individual, and no group of individuals that actually represent the nationality. The concept is an abstraction based on community of language which is felt by all as their mother tongue, and on the current forms of thought, feeling, and action—an abstraction of high emotional value, enhanced by the consciousness of political power, or by the desire for the power of independent control of the lives of the group.

Unity of language does not comprise the whole of nationalism, for no less ardent is the patriotism of trilingual Switzerland. Even here in America we see that the bond of tongue is not the only one. Else we should feel that there is no reason for a division between Canada and the United States or between the States of Spanish America, and that the political ties between western Canada and French Quebec must be artificial.

For the full development of his faculties, the individual needs the widest possible field in which to live and act according to his modes of thought and inner feeling. Since, in most cases, the opportunity is given among a group that possesses unity of speech, we feel full sympathy with the intense desire to throw down the artificial barriers of small political units. This process has characterized the development of modern nations.

When, however, these limits are overstepped, and a fictitious racial or alleged national unit is set up that has no existence in actual conditions, the free unfolding of the mind, for which we are striving, is liable to become an excuse for ambitious lust for power. The dream of a Pan-Latin Union, the Pan-Germanic wish for a union of all groups speaking Germanic languages, the Pan-Slavistic agitation, the Pan-American idea, are all prompted by the wish for power. A fictitious common culture and common racial origin is assumed on the basis of a relationship of language, discovered by philological research, but unrelated to modern culture. In all these cases the usefulness of the nationalistic idea was lost sight of and it was made the cover for the desire of imperialistic expansion.

The nationalism of modern times presupposes that the group held together as a nationality has developed the desire to strengthen its common social life, to determine its own actions, in other words, to become a nation which has the power to control its own destinies. Thus it has come to be the ferment that has broken up dynastic States comprising people that felt as distinct nationalities, and has led to the struggles for unity of those separated by the dynastic history of States.

The growth of modern, powerful States is the condition for the development of a strong nationalism.

Without a State conceived as an organization that can enforce and develop national aspirations nationality can never become the basis of a driving force.

This may readily be recognized when modern nationalism is compared with the intense group feelings of earlier periods. In small, tribal communities sameness of race, language and culture does not form a uniting tie. Each small social unit looks with suspicion, if not with enmity, upon its neighbors. It is conscious only of tribal solidarity. When a number of tribes form an organized confederation, like the Iroquois Indians, the community of interests and the centralization of social power in which all participate create a condition that may well be compared with modern nationalism. The question of linguistic unity is not determinant, but community of organization in peace and war. Among the Zulu in South Africa a rigid military organization created a national spirit, while the African States among which outlying districts were held to a central authority by looser bands had no stability and no national feeling.

Still more instructive is the absence of nationalistic feeling during the Middle Ages. In its place we find devotion to feudal lords and ruling dynasties. French battled against French, Italians against Italians according to their allegiance. While feudalism broke up the unity of what we should call nowadays a nationality, the unity of Christianity against Mohammedanism overstepped by far the limits of people of one speech. Both of these aspects of medieval life made impossible the feeling for a nationality as a uniting bond. The national State in our sense was non-existent.

It is well to bear in mind that nationalities may be created by a variety of circumstances. Economic interests and cultural contrasts may break a nation and create new nationalities. The break between the United States and England illustrates this. The feeling of national unity of the Southern States during the War of the Rebellion, created by community of economic interests and by the centralization of power made necessary by the war is another example.

An interesting phase of national life is developing in Russia. While the policy of the Czarist government consisted in the forcible suppression of all non-Russian speech, even of local dialects, the Soviet Republic has adopted the policy of protecting the right of every group to their own language, trusting the bond of a great, radical economic experiment to unite all the people as one nationality opposed to the capitalistic world.

There is no doubt that the idea of nationality has been a creative force, making possible the fuller development of powers by widening the field of individual activity, and by setting definite ideals to large coöperating masses; but we feel with Fichte and Mazzini that the political power of a nation is important only when the national unit is the carrier of ideals that are of value to mankind.

Together with the positive, creative side of nationalism there has developed everywhere an aggressive intolerance of foreign forms of thought that can be satisfied only by the strongest emphasis laid upon the value and interest of one national unit against all others.

On a larger scale the conditions are repeated now that less than a century ago prevented the ready formation of modern nations. The narrow-minded local interests of cities and other small political units resisted unification or federation on account of the supposed conflicts between their interests and ideals and those of other units of comparable size. Governmental organization strengthened the tendency to isolation, and the unavoidable, ever-present desire of self-preservation of the existing order stood in the way of amalgamation. It was only after long years of agitation and of bloody struggle that the larger idea prevailed.

Those of us who recognize in the realization of national ideals a definite advance that has benefited mankind cannot fail to see that the task before us at the present time is a repetition of the process of nationalization on a larger scale; not with a view to leveling down all local differences, but with the avowed purpose of making them all subserve the same end.

The federation of nations is the next necessary step in the evolution of mankind.

It is the expansion of the fundamental idea underlying the organization of the United States, of

Switzerland, and of Germany. The weakness of the League of Nations and of the modern peace movement lies in this, that they are not sufficiently clear and radical in their demands, for their logical aim cannot be arbitration of disagreement, or formal outlawing of war. It must be the recognition of common aims of all the nations.

Such federation of nations is not an Utopian idea, any more than nationalism was a century ago. In fact, the whole development of mankind shows that this condition is destined to come.

Fundamentally, the nation must be considered a closed society like those previously discussed. The differentiation between citizen and alien is not so intense as in the closed primitive horde, but it exists.

It would be instructive to follow in detail the development of modern nations from tribal units that considered every alien an enemy who must be slain, but we can only imagine the course of the gradual changes that have taken place.

Human inventions improved. The herd of hunters and food-gatherers learned the art of better providing for their needs. They stored up food and thus provided for the future. With the greater regularity of the food supply and a decreased frequency of periods of starvation the number of members of the community increased. Weaker hordes, which still followed the older methods of hunting and food gathering, were exterminated or, profiting by the experience of their neighbors, acquired new arts and also increased in numbers. Thus the groups that felt a solidarity among themselves became larger and by the extermination of small, isolated hordes, that remained in more primitive conditions, the total number of groups that stood opposed to one another became gradually less.

It is impossible to trace with any degree of certainty the steps by which the homogeneous groups became diversified and lost their unity, or by which the opposing groups came into closer contact. We may imagine that the widows and daughters of the slain, who became a welcome prey of the victors, established in time kindlier relations between their new masters and their kin; we may imagine that the economic advantages of peacefully acquiring the coveted property of neighbors rather than taking it by main force added their share to establishing kindlier relations; we may attribute an important influence to the weakening of old bonds of unity due to the gradual dispersion of the increasing number of members of the community. No matter how the next steps in political development happened, we see that, with increasing economic complexity, the hostility between the groups becomes less. If it was right before to slay every one outside of the small horde, we find now tribes that have a limited community of interests, that under normal conditions live at peace, although enmities may spring up at slight provocation. The group that lives normally at peace has much increased in size, and, while the feeling of solidarity may have decreased, its scope has become immensely wider.

This process of enlargement of political units and the reduction of the number of those that were naturally at war with one another began in the earliest times, and has continued without interruption, almost always in the same direction. Even though hostilities have broken out frequently between parts of what had come to be a large political unit, the tendency for unification has in the long run been more powerful than that of disintegration. We see the powers at work in antiquity, when the urban states of Greece and of Italy were gradually welded into larger wholes; we see it again at work after the breaking up of ancient society in the development of new states from the fragments of the old ones; and later on in the disappearance of the small feudal states.

In the nations of our days we find the greatest numbers of people united in political units that the world has seen. In these war is excluded, because all members are subject to the same law, and excessive strains in the community, that lead to internal bloodshed, have decreased in frequency, although perhaps not in violence, as long as the whole masses of the people in a nation enjoy somewhat equal advantages.

The World War has resulted in a setback to this movement that seems from our viewpoint as an anachronism. The breaking up of the old empire of Austria-Hungary is a step backward in a development that is steadily gaining in force. Notwithstanding the resistance of the governing class to the development of a confederation rather than of a centralized empire, the force of circumstances was operating in this direction. Hungary had attained a status of independence and the recognition of the rights of the South-Slavs was coming. How much better would the peacemakers have served humanity if they had created a confederacy of language groups of equal rights rather than a number of rival nations each of which is bent only upon the attainment of its own selfish ends!

Thus the history of mankind shows us the spectacle of the grouping of man in more or less firmly knit units of ever-increasing size that live together in peace, and that are ready to go to war only with other groups outside of their own limits. Notwithstanding all temporary revolutions and the shattering of larger units for the time being, the progress in the direction of recognition of common interests in larger groups, and consequent political federation has been so regular and so marked that we must needs conclude that the tendencies which have swayed this development in the past will govern our history in the future. The concept of thoroughly integrated nations of the size to which we are now accustomed would have been just as inconceivable in earlier times of the history of mankind as appears now the concept of unity of interests of all the peoples of the world, or at least of all those who share the same type of civilization and are subject to the same economic conditions. The historical development shows, however, that such a feeling of opposition of one group towards another is solely an expression of existing conditions, and does not by any means indicate their permanence.

It is not any rational cause that forms opposing groups, but solely the emotional appeal of an idea that holds together the members of each group and exalts their feeling of solidarity and greatness to such an extent that compromises with other groups become impossible. In this mental attitude we may readily recognize the survival of the feeling of specific differences between the hordes, transferred in part from the feeling of physical differences to that of mental differences. The modern enthusiasm for race superiority must be understood in this light. It is the old feeling of specific differences between social groups in a new guise.

Progress has been slow and halting in the direction of expanding the political units from hordes to tribes, from tribes to small states, confederations, and nations. The concept of the foreigner as a specifically distinct being has been so modified that we are beginning to see in him a member of mankind.

Enlargement of circles of association, and equalization of rights of distinct local communities have been so consistently the *general* tendency of human development that we may look forward confidently to their consummation.

It is obvious that the standards of ethical conduct must be quite distinct between those who have grasped this ideal and those who still believe in the preservation of the isolated nationality in opposition to all others.

Once we recognize this truth we are brought clearly face to face with those forces that will ultimately abolish warfare between nations as well as legislative conflicts; that will put an end not only to the wholesale slaughter of those representing distinct ideals, but also prevent the passage of laws that favor the members of one nation at the expense of all other members of mankind.

It should not be understood that such universalism is opposed to the development of individuality in nations. A large political unit may still be diverse in local culture and we should hesitate to foster any process that would bring us down to such a uniformity that the stimulus given by contact between different cultures should be lost, for contact between different attitudes and points of view has always been a force in keeping alive the intellectual and emotional activities of mankind.

In primitive society existed an immense variety of cultural forms in contiguous areas. Isolation prevented leveling down of differences, although a trickling through of cultural streams may be observed. We have lost much of this diversity, but local characteristics of culture persist, expressed in emotional attitudes, forms of social intercourse, intellectual interests and occupations, in the valuation of the character and activities of man.

Much fuller developed than in primitive society are the differences in cultural outlook of the various strata of civilized society. Notwithstanding the sameness of the products of the civilization in which we all live, fundamental differences are found, and when the isolation between the strata comes to be great or the contrasts are accompanied by economic distress in some of the strata, forcible oppression or revolt results.

The suppression of cultural differences or isolation of the different groups cannot be the aim of intelligent endeavor in directing human development.

However, in our educational systems cultural nationalism is hardly mentioned, political nationalism is stressed. Devotion to the political interests of the nation, to political power, is taught as the paramount duty and is instilled into the minds of the young in such a form that with it grows up and is perpetuated the feeling of rivalry and of hostility against all other nations.

Conditions in modern states are intelligible only when we remember that through education patriotism is surrounded with a halo of sanctity and that national self-preservation is considered the first duty. Often the demands of national and international duty are hopelessly at variance.

The interests of mankind are ill served if we try to instill into the minds of the young a passionate desire for national power; if we teach the preponderance of national interest over human interest, aggressive nationalism rather than national idealism, expansion rather than inner development, admiration of warlike, heroic deeds rather than of the object for which they are performed.

CHAPTER V

EUGENICS

THE POSSIBILITY of raising the standards of human physique and mentality by judicious means has been preached for years by the apostles of eugenics, and has taken hold of the public mind to such an extent that eugenic measures have found a place on the statute books of a number of States, and that the public conscience disapproves of marriages that are thought bound to produce unhealthy offspring.

The thought that it may be possible by these means to eliminate suffering and to strive for higher ideals is a beautiful one, and makes a strong appeal to those who have at heart the advance of humanity.

Our experiences in stock and plant breeding have shown that it is feasible, by appropriate selection, to change a breed in almost any direction that we may choose: in size, form, color. Even physiological functions may be modified. Fertility may be increased, speed of movement improved, the sensitiveness of sense organs modified, and mental traits may be turned in special directions. It is, therefore, more than probable that similar results might be obtained in man by careful mating of appropriately selected individuals,—provided that man allowed himself to be selected in the same manner as we select animals. We have also the right to assume that, by preventing the propagation of mentally or physically inferior strains, the gross average standing of a population may be raised.

Although these methods sound attractive, there are serious limitations to their applicability. Eugenic selection can affect only hereditary features. If an individual possesses a desirable quality the development of which is wholly due to environmental causes, and that will not be repeated in the descendants, its selection will have no influence upon the following generations. It is, therefore, of fundamental importance to know what is hereditary and what not. Features, and color of eyes, hair and skin, are more or less rigidly hereditary; in other words, in these respects children resemble organically their parents, no matter in what environment they may have been brought up. In other cases, however, the determining influence of heredity is not so clear. We know that stature depends upon hereditary causes, but that it is also greatly influenced by environmental conditions prevailing during the period of growth. Rapidity of development is no less influenced by these two causes, and in general the more subject an anatomical or physiological trait to the influence of environment the less definitely can we speak of a controlling influence of heredity, and the less are we justified in claiming that nature, not nurture, is the deciding element.

It would seem, therefore, that the first duty of the eugenist should be to determine empirically and without bias what features are hereditary and what not.

Unfortunately this has not been the method pursued; but the battle cry of the eugenists, "Nature not nurture," has been raised to the rank of a dogma, and the environmental conditions that make and unmake man, physically and mentally, have been relegated to the background.

It is easy to see that in many cases environmental causes may convey the erroneous impression of hereditary phenomena. Poor people develop slowly and remain short of stature as compared to wealthy people. We find, therefore, in a poor area, apparently a low hereditary stature, that, however, would change if the economic life of the people were changed. We find proportions of the body determined by occupations, and apparently transmitted from father to son, provided both father and son follow the same occupation. The more far-reaching the environmental influences are that act upon successive generations the more readily will a false impression of heredity be given.

Here we reach a parting of the ways of the biological eugenist and the student of human society. Most modern biologists are so entirely dominated by the notion that function depends upon form that they seek for an anatomical basis for all differences of function. The stress laid upon the relation between anatomical form or constitution and pathological conditions of the most varied character is an expression of this tendency. Whenever the anatomical and pathological conditions are actually physiologically interdependent such relations are found. In other cases, as for instance in the relation of anatomical form and mental disturbances, the relation may be quite remote. This is still more the case when a relation between social phenomena and bodily form is sought. Many biologists are inclined to assume that higher civilization is due to a higher type; that better social health depends solely upon a better hereditary stock; that national characteristics are determined by the bodily forms represented in the nation.

The anthropologist is convinced that many different anatomical forms can be adapted to the same social functions; and he ascribes greater weight to these and believes that in many cases differences of form may be due to adaptations to different functions. He believes that different types of man may reach the same civilization, that for any type of man better health may be secured by better nurture.

The anatomical differences and those in chemical constitution to which the biologist reduces social phenomena are hereditary; the environmental causes which the anthropologist sees reflected in human form are individually acquired, and not transmitted by heredity.

In view of what has been said before it will suffice to point out a very few examples.

Sameness of language is acquired under the same linguistic environment by members of the most diverse human types; the same kinds of foods are selected from among the products of nature by people belonging to the same cultural area; similarity of movements is required in industrial pursuits; the habits of sedentary or nomadic life do not depend upon race but upon occupation. All of these are distributed without any reference to physical type, and give ample evidence of the lack of relation between social habits and racial position.

The serious demand must be made that eugenists cease to look at the forms, functions, and activities of man from the dogmatic point of view according to which each feature is assumed to be hereditary, but that they begin to examine them from a more critical point of view, requiring that in each and every case the hereditary character of a trait must be established before it can be assumed to exist.

The question at issue is well illustrated by the extended statistics of cacogenics, of the histories of defective families. Setting aside for a moment cases of hereditary pathological conditions, we find that alcoholism and criminality are particularly ascribed to hereditary causes. When we study the family histories in question, we can see often, that, if the individuals had been protected by favorable home surroundings and by possession of adequate means of support against the abuse of alcohol or other drugs as well as against criminality, many of them would have been no more likely to fall victims to their alleged hereditary tendencies, than many a weakling who is brought up under favorable circumstances. If they had resisted the temptations of their environment they would have been entitled to be classed as moral heroes. The scales applied to the criminal family and to the well-to-do are clearly quite distinct; and, so far as heredity is concerned, not much more follows from the collected data of social deficiencies than would follow from the fact that in an agricultural community the occupation of farmers descends from father to son.

Whether or not constitutional debility based on hereditary causes may also be proved in these cases is a question by itself that deserves attention. It remains to be proved in how far it exists, and furthermore it cannot be assumed without proof that the elimination of the descendants of delinquents would free us of all those who possess equal constitutional debility. Of these matters more anon.

It is an observed fact that the most diverse types of man may adapt themselves to the same forms of life and, unless the contrary can be proved, we must assume that all complex activities are socially determined, not hereditary; that a change in social conditions will change the whole character of social activities without influencing in the least the hereditary characteristics of the group of individuals concerned.

Therefore, when the attempt is made to prove that defects or points of excellence are hereditary, it is essential that all possibility of a purely environmentally or socially determined repetition of ancestral traits be excluded.

If this rigidity of proof is insisted on it will appear that many of the data on which the theory of eugenics is based are unsatisfactory, and that much greater care must be exerted than finds favor with the enthusiastic adherents of eugenic theories.

All this does not contradict the hereditary transmission of individual physical and mental characteristics, or the possibility of segregating, by proper selection from among the large series of varying individual forms that occur among all types of people, strains that have admirable qualities, and of suppressing others that are not so favored.

It is claimed that the practical application has become a necessity because among all civilized nations there is a decided tendency to general degeneration. I do not believe that this assertion has been adequately proved. In modern society the conditions of life have become markedly varied as compared with those of former periods. While some groups live under most favorable conditions, that require active use of body and mind, others live in abject poverty, and their activities have more than ever before been degraded to those of machines. At the same time, human activities are much more varied than formerly. It is, therefore, quite intelligible that the functional activities of each nation must show an increased degree of differentiation, a higher degree of variability. The general average of the mental and physical types of the people may remain the same, still there will be a larger number now than formerly who fall below a certain given low standard, while there will also be more who exceed a given high standard. The number of defectives can be counted by statistics of poor relief, delinquency and insanity, but there is no way of determining the increase of those individuals who are raised above the norm of a higher standard. Therefore they escape our notice. It may very well be that the number of defectives increases, without, however, influencing the value of a population as a whole, because it is merely an expression of an increased degree of variability.

Furthermore, arbitrarily selected, absolute standards do not retain their significance. Even if no change in the absolute standards should be made, the degree of physical and mental energy required under modern conditions to keep one's self above a certain minimum of achievement is higher than formerly. This is due to the greater complexity of our life and to the increasing number of competing individuals. When the general level of achievement is raised, greater capacity is required of those who are to attain a high degree of prominence than was needed in earlier periods óf our history. A mentally defective person may be able to hold his own in a simple farming community and unable to do so in city life. The claim that we have to contend against national degeneracy must, therefore, be better substantiated than it is now.

The problem is further complicated by the advances of public hygiene, which have lowered infant mortality, and have changed the composition of the population, in so far as many who would have succumbed to deleterious conditions in early years enter into the adult population and have an influence upon the general distribution of vitality.

There is still another important aspect of eugenics that should make us pause before we accept this new ambitious theory as a panacea for human ills. The radical eugenist treats the problem of procreation from a purely rationalistic point of view, and assumes that the ideal of human development lies in the complete rationalization of human life. As a matter of fact, the conclusions to be drawn from the study of the customs and habits of mankind show that such an ideal is unattainable, and more particularly that the emotions clustering about procreation belong to those that are most deeply seated, and that are ineradicable.

Here again the anthropologist and the biologist are at odds. The natural sciences do not recognize in their scheme a valuation of the phenomena of nature, nor do they count emotions as moving forces; they endeavor to reduce all happenings to the actions of physical causes. Reason alone reigns in their domain. Therefore the scientist likes to look at mental life from the same rational standpoint, and sees as the goal of human development an era of reason, as opposed to the former periods of unhealthy fantastic emotion.

The anthropologist, on the other hand, cannot acknowledge such a complete domination of emotion by reason. He rather sees the steady advance of the rational knowledge of mankind, which is a source of satisfaction to him no less than to the biologist; but he sees also that mankind does not put this knowledge to purely reasonable use, but that its actions are swayed by emotions no less now than in former times, although in many respects, unless the passions are excited, the increase of knowledge limits the extreme forms of unreasonable emotional activities. Religion and political life, and our everyday habits, present endless proofs of the fact that our actions are the results of emotional preferences, that conform in a general way to our rational knowledge, but which are not determined by reason; that we rather try to justify our choice of action by reason than have our actions dictated by reason.

It is, therefore, exceedingly unlikely that a rational control of one of the strongest passions of man could ever succeed. If even in matters of minor importance evasion of the law is of common occurrence, this would be infinitely more common in questions that touch our inner life so deeply. The repugnance against eugenic legislation is based on this feeling.

It cannot be doubted that the enforcement of eugenic legislation would have a far-reaching effect upon social life, and that it would tend to raise the standard of certain selected hereditary strains. It is, however, an open question what would happen to the selected strains owing to the changed social ideals; and it is inexcusable to refuse to consider those fundamental changes that would certainly be connected with eugenic practice, and to confine ourselves to the biological effect that may be wrought, for in the great mass of a healthy population the biological mechanism alone does not control social activities. These are rather subject to social stimuli.

Although we are ignorant of the results of a rigid application of eugenics, a few of its results may be foretold with great certainty.

The eugenist who tries to do more than to eliminate the unfit will first of all be called upon to answer the question what strains are the best to cultivate. If it is a question of breeding chickens or Indian corn, we know what we want. We desire many eggs of heavy weight, or a large yield of good corn. But what do we want in man? Is it physical excellence, mental ability, creative power, or artistic genius? We must select certain ideals that we want to raise. Considering then the fundamental differences in ideals of distinct types of civilization, have we a right to give to our modern ideals the stamp of finality, and suppress what does not fit into our life? There is little doubt that we, at the present time, give much less weight to beauty than to logic. Shall we then try to raise a generation of logical thinkers, suppress those whose emotional life is vigorous, and try to bring it about that reason shall reign supreme, and that human activities shall be performed with clocklike precision? The precise cultural forms that would develop cannot be foretold, because they are culturally, not biologically, determined; but there is little doubt that within certain limits the intensity of emotional life,—regardless of its form,—and the vigor of logical thought,—regardless of its content,—could be increased or decreased by organic selection. Such a deliberate choice of qualities which would modify the character of nations implies an overestimation of the standards that we have reached, which to my mind appears intolerable. Personally the logical thinker may be most congenial to me, nevertheless I respect the sacred ideal of the dreamer who lives in a world of musical tones, and whose creative power is to me a marvel that surpasses understanding.

Without a selection of standards, eugenic practice is impossible; but if we read the history of mankind aright, we ought to hesitate before we try to set our standards for all time to come, for they are only one phase in the development of mankind.

This consideration applies only to our right to apply creative eugenic principles, not to the question whether practical results by eugenic selection can be attained. I have pointed out before how much in this respect is still hypothetical, or at least of doubtful value, because the social factors outweigh the biological ones.

At the present time the idea of creating the best human types by selective mating is hardly a practical one. It dwells only as a desirable ideal in the minds of some enthusiasts.

The immediate application of eugenics is rather concerned with eliminating strains that are a burden to the nation or to themselves, and with raising the standard of humanity by the suppression of the progeny of the defective classes. I am doubtful whether eugenics alone will have material results in this direction, for, in view of the fundamental influence of environmental causes, that I set forth before, it is perfectly safe to say that no amount of eugenic selection will overcome those social conditions that have raised a poverty- and disease-stricken proletariat—which will be reborn from even the best stock, so long as the social conditions persist that remorselessly push human beings into helpless and hopeless misery. The effect would probably be to push new groups of individuals into the deadly environment where they would take the place of the eliminated defectives. Whether they would breed new generations of defectives may be an open question. The continued presence of defectives would be a certainty. Eugenics alone cannot solve the problem. It requires much more an amelioration of the social conditions of the poor which would also raise many of the apparently defective to higher levels.

The present state of our knowledge of heredity permits us to say that certain pathological conditions are hereditary and that apparently healthy parents who belong to defective strains are very likely to have among their descendants defective individuals. We may even predict for a number of such cases how many among the descendants will be normal and how many defective. The eugenist must decide whether he wants to suppress all the normal individuals in these families in order to avoid the development of the defectives, or whether he is willing to carry the defectives along, perhaps as a burden to society, to their relatives and in many cases even to themselves, for the sake of the healthy children of such families. This question cannot be decided from a scientific point of view. The answer depends upon ethical and social standards. Many defective families have produced individuals who have given us the greatest treasures our civilization possesses. Eugenists might have prevented Beethoven's father from having children. Would they willingly take the responsibility of having mankind deprived of the genius of Beethoven?

Another aspect of the problem is of much more vital importance to mankind. The object of eugenics is to raise a better race and to do away with increasing suffering by eliminating those who are by heredity destined to suffer and to cause suffering. The humanitarian idea of the conquest of suffering, and the ideal of raising human efficiency to heights never before reached, make eugenics particularly attractive.

I believe that the human mind and body are so

constituted that the attainment of these ends would lead to the destruction of society. The wish for the elimination of unnecessary suffering is divided by a narrow margin from the wish for the elimination of all suffering.

While, humanely speaking, this may be a beautiful ideal, it is unattainable. The performance of the labors of mankind and the conflict of duties will always be accompanied by suffering that must be borne, and that men must be willing to bear. Many of the works of sublime beauty are the precious fruit of mental agony; and we should be poor, indeed, if the willingness of man to suffer should disappear. However, if we cultivate this ideal, then that which was discomfort yesterday will be suffering to-day, and the elimination of discomforts will lead to an effeminacy that must be disastrous to the race.

This effect is further emphasized by the increasing demands for self-perfection. The more complex our civilization and the more extended our technical skill and our knowledge, the more energy is demanded for reaching the highest efficiency, and the less is it admissible that the working capacity of the individual should be diminished by suffering. We are clearly drifting towards that danger-line where the individual will no longer bear discomfort or pain for the sake of the continuance of the race, and where our emotional life is so strongly repressed by the desire for self-perfection—or by self-indulgence—that the coming generation is sacrificed to the selfishness of the living, and the more so the more competent each one to make use of his natural gifts. The phenomenon that characterized the end of antiquity, when no children were born to take the place of the passing generations, is being repeated in our times and in ever widening circles; and the more vigorously the eugenic ideals of the elimination of suffering and of self-development are held up the sooner shall we drift towards the destruction of the race.

Eugenics should, therefore, not be allowed to deceive us into the belief that we should try to raise a race of supermen, nor that it should be our aim to eliminate all suffering and pain. The attempt to suppress those defective classes whose deficiencies can be proved by rigid methods to be due to hereditary causes, and to prevent unions that will unavoidably lead to the birth of disease-stricken progeny, is the proper field of eugenics. How much can be and should be attempted in this field depends upon the results of careful studies of the laws of heredity. Eugenics is not a panacea that will cure human ills; it is rather a dangerous sword that may turn its edge against those who rely on its strength.

CHAPTER VI

CRIMINOLOGY

A WHOLE science has developed based on the assumption of the existence of a biologically determined criminal type and upon the hereditary transmission of criminality. The Italian school of criminologists led by C. Lombroso has endeavored to define the type of the criminal and the physical characteristics of criminals addicted to various types of crimes. A number of stigmata have been established which, it was believed, characterized a person as a criminal. If this theory could be proved the treatment of criminals would have been much simplified, for it would have been possible to select all criminals before the commission of a crime and to protect society against them.

Unfortunately these extreme hopes have not been fulfilled. Our previous considerations make it plausible that they could not be fulfilled, because the interrelation between gross bodily form and mentality is not by any means close.

All that has been proved is that many criminals are defective, not only mentally but also physically.

It is, therefore, not surprising that anomalies that accompany various types of defectiveness should be found among them with greater frequency than among the socially normal; but it does not follow that the presence of any one of the stigmata described by the Italian school would prove that a person is a born criminal.

In many of the cases a careful statistical study has shown that the alleged stigmata, such as absence of the lobe of the ear and irregularities in the position of the teeth, are more frequent in local noncriminal groups than among the criminals, so that for this reason they cannot be considered as significant. Neither is there any clear physiological relation between the alleged stigmata and social or even physical defects.

A most careful examination of the criminal population has been made by C. Goring. His general results are worth quoting. He says: "For statistical evidence, one assertion can be dogmatically made: It is, that the criminal is differentiated by inferior stature, by defective intelligence and, to some extent, by his antisocial proclivities; but that, apart from these broad differences, there are no physical, mental or moral characteristics peculiar to the inmates of English prisons. The truths that have been overlooked are that these deviations, described as significant of criminality, are inevitable concomitants of inferior stature and defective intelligence: both of which are the differentia of the types of persons who are selected for imprisonment."

The conditions are the same as those previously described. As it is impossible to assign an individual according to his bodily form to a racial group, if the groups overlap, so it is impossible to recognize an individual by his bodily build as a criminal. We may say that it is more likely that a person physically and mentally defective will become a criminal than one who is normal, but we cannot say that he *must* be a criminal.

The very definition of the term "crime" proves that no such intimate relation can exist. What was a crime in times past is no longer a crime now. Heresy was a crime punishable by death. Among heretics were included many who were mentally unbalanced and probably physically defective; but men like Huss or Giordano Bruno were criminals on account of their mental independence. George Washington would have been a criminal, if the English had caught him.

In foreign societies the concept of what constitutes a crime may be even more different than it has been at different periods among ourselves. Where food is shared by all and property consists solely of the necessities of life, such as clothing, weapons, household utensils, small pilfering is all but impossible, for the taking of food is not stealing, food being freely shared by all. Where strict laws of endogamy exist, what we call incest may be prescribed. Where exogamy is found the laws of incest extend over wider, or curiously selected groups. Where vendetta is the law of the land certain types of murder are a virtue, not a crime. Where monogamy is the custom polygamy is a criminal offense, while in other societies the refusal to accept a number of mates may be so considered. Where sexual life is practically free sexual crimes do not occur.

Under these conditions the criminal must be defined as the person who habitually disregards the laws of conduct prescribed by the society to which he belongs. If we accept this definition we must except those cases in which conduct contrary to law is ceremonially permitted or prescribed. This happens, for instance, among the Pueblo Indians and in British Columbia in the case of certain semi-priestly groups who have the privilege of acting counter to the sacred rites of the people and who are accordingly feared by the profane crowd. The same is true in all cases of prerogatives of social classes—as in the relation between master and slave, when the slave is considered a chattel; or in prerogatives of feudal lords.

With the differentiation of what constitutes a crime the mental characteristics of the criminal must also vary. The criminal who breaks through the inhibitions developed by the habitual behavior of the society to which he belongs is actuated by a variety of motives. The breaking point depends upon the drive that leads to action and the strength of inhibition. Among two persons with equal power of inhibition the starving pauper will be led to theft by hunger; the well-to-do who is deprived of his conveniences will succumb much more readily, because the strain which for the pauper would be insignificant is felt by him as suffering. Such conditions may account for the similar distribution of criminality in well-to-do and poor social groups.

The problem of the hereditary determination of criminality as well as of other forms of social deficiency presents the same difficulties that are encountered in all attempts to discriminate between organic and environmental determination.

The definition of crime is so complex and so variable, so entirely dependent upon social conditions that criminality itself can hardly be considered as hereditary. It is, however, possible that certain dispositions may be hereditary that lead to acts that are in some cases considered as criminal. It has been proved that the criminal is, in many respects, defective. If the deficiency is hereditary, then a greater probability exists that a defective individual belonging to a hereditary line of defectives may become a criminal.

The investigation of families like the Kallikaks has shown that there are strains in which criminality is very frequent. From a purely practical point of view these data allow us to say that when a person is a criminal or otherwise defective there is a greater likelihood of finding criminals or defectives in his family than among the relatives of a person who is not a criminal.

The reason for this is easily understood if we remember that the same is true for any trait that occurs comparatively rarely and with unequal frequency in different families. If in a preponderantly blond population a blond is selected we exclude all those families in which no blonds occur and the average frequency of blondness in the population thus selected will be considerable. On the other hand, if we select a brunette individual the whole mass of families that contain brunette individuals will appear, and the average frequency of blondness in the series thus selected will be much lower. The same is true when we select exceptionally short individuals. Then all tall families will be eliminated the more the higher their average stature, and the series so selected will contain an inordinately large number of short individuals. Conversely in the series of families selected as relatives of a tall person the relative frequency of short ones will be much less. The statistical value obtained from such data will depend entirely upon the frequencies with which blondness and tallness appear in all the families investigated. If one group of families had only blond or tall individuals, the others none, all the members of the families in which one tall person occurs will be tall. If some families have many individuals of a specific trait, others few, we cannot be certain how many of the relatives of a person characterized by a given trait will share it and the law of hereditary transmission cannot be established. Therefore the separation of hereditary determination or acquisition through external conditions cannot be made.

Tallness and shortness are not entirely due to heredity, so that from the greater frequency of short relatives of a short man we may not immediately conclude that shortness is hereditary.

This is still clearer in social groups. In rural communities the relatives of a farmer are preponderantly farmers, but farming is not a hereditary trait. The relatives of a prosperous business man are rarely unskilled laborers.

In short, the frequency of any trait in a family line does not, without further proof, show that it is hereditary. The greater frequency of criminality among relatives of criminals does not allow us to deduce laws of the heredity of criminality, unless the hereditary determination is proved by other methods.

We have seen that the family lines constituting a population differ among themselves. They differ also in regard to criminality and frequency of defects. The questions to be answered are whether these are environmentally determined or hereditary and what the laws of heredity are.

The observations of Habit-Clinics for pre-school children throw an interesting light upon this problem. Although the statistical results of these observations must be used with considerable caution, the psychological analysis elucidates the far-reaching influence of an unfavorable environment upon the behavior of physically weak subjects and the development of antisocial tendencies that may arise under stresses of a family situation that make for revolt against tyrannical authority or create in other ways serious antagonisms.

No less instructive are the observations of psychoanalysis. While I am not inclined to follow the intricate and, as it seems to me, arbitrary reasonings of psychoanalysts, sufficient material has been accumulated showing that under severe stresses, particularly after a sudden "trauma," weak individuals may develop abnormal mental habits of the most varied kind.

The general evidence points to the conclusion that the weak individual takes to antisocial acts when the environmental stress that brings about disregard of the laws of society is sufficiently acute. The stronger the individual the greater the stress that will be required.

C. Goring, in the investigation previously referred to, minimizes the environmental factor as a determinant of criminality. He tries to prove that all other social irregularities found among criminals, such as lack of schooling or irregular employment, or poverty are dependent upon lack of intelligence. His argument is based on the statistical interrelation between intelligence and the various social defects. He determines the average intelligence of a group by the relative frequency of mental defectives. He assumes that the greater their number the lower the average intelligence. This is a doubtful procedure, because the range of variation in the groups does not need to be the same. If, for instance, the mentality of criminals were more variable than that of noncriminals, they would have a larger number of defectives even if they had the same average intelligence. Social irregularities combined with criminality are the more frequent the greater the relative number of mental defectives. The argument might also be reversed and we might say that mental defects combined with criminality are the more frequent the greater the relative number of social irregularities, such as lack of schooling or irregular employment. In order to prove that organically determined intelligence is the cause of both social irregularities and criminality it would be necessary to show that groups of individuals of the same intelligence, taken at random from the total population, would have the same relative frequency of criminality regardless of other social defects, such as poverty, lack of schooling or irregularity of employment. Since we do not know the distribution of intelligence in the total population the ratio of criminality cannot be determined and it cannot be claimed that hereditary intelligence is the decisive factor.

I believe, therefore, that the irrelevancy of environment as a factor producing criminality has not been proved.

Many authors have tried to deduce from the distribution of cases of criminality in family lines that the tendency is inherited in a simple Mendelian ratio. The infinite complexity of conditions that bring an individual into the class of convicted criminals does not make such a conclusion likely and the number of cases that have been brought forward is entirely insufficient for a conclusive proof. The actual statistical data indicate only that in the population family lines differ in their degree of criminality.

The assumption of a simple form of Mendelian heredity, and that of the occurrence of much more complex forms which include environmental factors lead to quite distinct practical results. In the former case the occurrence of a single case of criminality in a family and a knowledge of the simple rules of hereditary transmission would enable us to foretell how many individuals in various family lines would be affected. In the latter case prediction would be well-nigh impossible, because the rules of heredity, although following fundamentally the same laws, would be so varied that the hereditary characteristics of a single family would not be known.

More important than this is the difficulty of differentiation between environmental and hereditary causes, for if a whole family is exposed to the same deleterious conditions and a sufficient organic weakness exists, the whole family may become criminal, while under more fortunate conditions it could withstand the social pressure to which it is exposed.

CHAPTER VII

STABILITY OF CULTURE

AN ISOLATED community that remains subject to the same environmental conditions, and without selective mating, becomes, after a number of generations, stable in bodily form. As long as there are no stimuli that modify the social structure and mental life the culture will also be fairly permanent. Primitive, isolated tribes appear to us and to themselves as stable, because under undisturbed conditions the processes of change of culture are slow.

In the very earliest times of mankind culture must have changed almost imperceptibly. The history of man, of a being that made tools, goes back maybe 150,000 years, more or less. The tools belonging to this period are found buried in the soil. They are stone implements of simple form. For a period of no less than 30,000 years the forms did not change. When we observe such permanence among animals we explain it as an expression of instinct. Objectively the toolmaking of man of this period seems like an instinctive trait similar to the instincts of ants and bees. The repetition of the same act without change, generation after generation, gives the impression of a biologically determined instinct. Still, we do not know that such a view would be correct, because we cannot tell in how far each generation learned from its predecessors. Animals like birds and mammals, act not only instinctively; they also learn by example and imitation. Horses and dogs learn to react to calls or to the spoken word. English sparrows reared by canaries learn their song and call-notes. Parrots learn to imitate sounds. Apes even learn to use sticks or stones as tools.

It seems likely that conditions were the same in early man. Even in the earliest remains differences may be found. While in some areas the typical form of an implement was the flint blade, in others it was the cleaver or coup-de-point. According to Menghin a culture based on the use of bone originated in arctic Asia, another one based on the manufacture of flint blades in Eastern Asia, and one based on the flint cleaver in India.

The importance of the process of learning becomes more and more evident the nearer we approach the present period. The tools become more differentiated. Not all localities show the same forms, and it seems likely that if we could examine the behavior of man in periods one thousand years apart that changes would be discovered.

At the end of the ice age the differentiation in the forms of manufactured objects had come to be as great as that found nowadays among primitive tribes.

There is no reason why we should assume the life of the people who lived towards the end of the ice age, the Magdalenians, to have been in any respect simpler than that of the modern Eskimo.

With the beginning of the present geological period the differentiation of local groups and of activities in each group was considerable. Changes which in the beginning required tens of thousands of years, later thousands of years, occurred now in centuries and brought about constantly increasing multiplicity of forms.

With the approach of the historic period the degree of stability of culture decreased still further and in modern times changes are proceeding with great rapidity, not only in material products of our civilization but also in forms of thought.

Since earliest times the rapidity of change has grown at an ever-increasing rate.

The rate of change in culture is by no means uniform. We may observe in many instances periods of comparative stability followed by others of rapid modifications. The great Teutonic migrations at the close of antiquity brought about fundamental changes in culture and speech. They were followed by periods of consolidation. The Arab conquest of North Africa destroyed an old civilization and new forms took its place. Assimilation of culture may also be observed among many primitive tribes, and, although we do not know the rate of change, there is often strong internal evidence of a rapid adjustment to a new level. In language the alternation between periods of rapid change and comparative stability may often be observed. The transition from Anglo-Saxon and Norman to English was rapid. The development of English since that time has been rather slow. Similar periods of disturbance have occurred in the development of modern Persian.

Changes of unusual rapidity are due to the influence of European civilization upon primitive cultures. When they do not completely disappear a new adjustment is reached with great rapidity. This is exemplified by the modern culture of the Indians of Mexico and Peru. Part of their ancient material culture survives. Under the veneer of catholicism and of other Spanish cultural forms old ideas persist, readjusted to the superimposed civilization. A blend has developed which does not yield until modern schools and a livelier participation in world affairs disturb the equilibrium. A remarkable example of adjustment between old and new is found among the Pueblo tribes of New Mexico who have consciously and as far as possible isolated themselves from the American life around them. Their daily life has been modified by the use of products of American manufacture. Woven goods, glass windows and doors, agricultural implements, household furniture are in use; Catholic churches are attended on Sundays; the Saints' days are celebrated; and all this is assimilated to the older forms of life. The ancient house forms persist; in some Pueblos the former style of dress survives; as heretofore, corn is ground on the grinding stone; old types of Spanish ovens for baking bread continue to be used, and the ancient religious beliefs and ceremonials have been so adjusted that they continue, without serious inner conflicts, side by side with catholicism. The new equilibrium is disturbed only when the general conditions of life make continued isolation impossible and the younger generation finds a new adjustment to altered conditions.

Even more striking is the rapidity of change of culture among the Negroes of the United States. Since their introduction as slaves their language, their ancient customs and beliefs, have disappeared apace with their absorption in the economic life of America. Dr. Parsons, Dr. Herskovits, and Miss Zora Hurston have shown that, as we proceed from south to north, from Dutch Guyana to the northern States, the survivals of Negro culture become less and less. The isolated Bush Negroes of Surinam are essentially African in culture. The Negro districts of the South retain some African elements, while the northern Negro city dweller is to all intents and purposes like his White neighbor, except in so far as social barriers tend to perpetuate one or the other peculiarity of behavior.

Notwithstanding the rapid changes in many aspects of our modern life we may observe in other respects a marked stability. Characteristics of our civilization are conflicts between the inertia of conservative tradition and the radicalism which has no respect for the past but attempts to reconstruct the future on the basis of rational considerations intended to further its ideals. These conflicts may be observed in education, law, economic theory, religion, and art. Discipline against freedom of control, subordination under the public weal against individual freedom, capitalism against socialism, dogma against freedom of belief, established art forms against esthetic expression subject only to individual whim, are some of these conflicts. They are possible only when in a rapidly changing culture the old and the new live side by side.

We are wont to measure the ability of a race by its cultural achievements which imply rapid changes. Those races among whom the later changes have been most rapid appear, therefore, as most highly developed.

For these reasons it is important to study the conditions that make for stability and for change; and to know whether changes are organically or culturally determined.

Behavior that is organically determined is called instinctive. When the infant cries and smiles, when later on it walks, its actions are instinctive in this sense. Breathing, chewing, retiring from a sudden assault against the senses, approach towards desired objects are presumably organically determined. They do not need to be learned. Most of these actions are indispensable for the maintenance of life. We can never account for the reasons that prompt us to perform acts organically determined. The stimulus presents itself and we react at once, without conscious effort. Still, some of these reactions may be modified or even suppressed with impunity. Thus we may learn to overcome the reaction to fear. It is difficult to do so, but not impossible.

On the ground of this experience we are inclined to consider every type of behavior that is marked by an immediate, involuntary reaction as instinctive. This is an error, for habits imposed upon us during infancy and childhood have the same characteristics. They determine the particular forms of our activities, even of those based on the structure of our organism. We must recognize that the specific *forms* of our actions are culturally determined.

We must eat in order to live. Arctic man is compelled by necessity to live on a meat diet; the Hindu lives on vegetal food by choice.

That we walk on our legs is organically conditioned. How we walk, our particular gait, depends upon the forms of our shoes, the cut of our clothing, the way we carry loads, the conformation of the ground we tread. Peculiar forms of motion may be, in part, physiologically determined, but many are due to imitation. They are repeated so often that they become automatic. They come to be the way in which we move "naturally." The response is as easy and as ready as an instinctive action, and a change from the acquired habit to a new one is equally difficult. When thoroughly established the level of consciousness of an automatic action is the same as that of an instinctive reaction.

In all these cases the *faculty* of developing a certain motor habit is organically determined. The particular *form* of movement is automatic, acquired by constant, habitual use.

This distinction is particularly clear in the use of language. The *faculty* of speech is organically determined and should be called, therefore, instinctive. However, *what* we speak is determined solely by our environment. We acquire one language or another, according to what we hear spoken around us. We become accustomed to very definite movements of lips, tongue and the whole group of articulating organs. When we speak, we are wholly unconscious of any of these movements and equally of the structure of the language we speak. We resent deviations in pronunciation and in structure. As adults we find it exceedingly difficult, if not impossible, to acquire complete mastery of new articulations and new structures such as are required in learning a foreign language. Our linguistic habits are not instinctive. They are automatic.

Our thoughts and our speech are accompanied by muscular movements—some people would even say they *are* our thoughts. The kinds of movements are not by any means the same everywhere. The mobility of the Italian contrasts strikingly with the restraint of the Englishman.

The human faculty of using tools is organically determined. It is instinctive. This, however, does not mean that the kind of tool developed is prescribed by instinct. Even the slightest knowledge of the development of tools proves that the special forms characteristic of each area and period depend upon tradition and are in no way organically determined. The choice of material depends partly upon environment, partly upon the state of inventions. We use steel and other artificially made materials; the African iron, others stone, bone, or shell. The forms of the working parts of the implements depend upon the tasks they are to perform, those of the handles upon our motor habits.

The same is ordinarily true of our likes and dislikes. We are organically capable of producing and enjoying music. What kind of music we enjoy depends for most of us solely upon habit. Our harmonies, rhythms, and melodies are not of the same kind as those enjoyed by the Siamese and a mutual understanding, if it can be attained at all, can be reached solely by long training.

Whatever is acquired in infancy and childhood by unvarying habits becomes automatic.

There is a negative effect of automatism, no less important than the positive one which results in the ease of performance.

Any action that differs from those performed by us habitually strikes us immediately as ridiculous or objectionable, according to the emotional tone that accompanies it. Often deviations from automatic actions are strongly resented. A dog taught to give his hind paw instead of the front paw excites us to laughter. Formal dress worn at times when the conventions do not allow it seems ridiculous. So does the dress that was once fashionable but that has gone out of use. We need only think of the hoop skirt of the middle of the last century or of the bright colors of man's dress and the impression they would create to-day. We must also realize the resistance that we ourselves have to appearing in an inappropriate costume.

More serious are the resistances in matters that evoke stronger emotional reactions. Table manners are a good example. Most of us are exceedingly sensitive to a breach of good table manners. There are many tribes and people that do not know the use of the fork and who dip into the dish with their fingers. We feel this is disgusting because we are accustomed to the use of fork and knife. We are accustomed to eat quietly. Among some Indian tribes it is discourteous not to smack one's lips, the sign of enjoying one's food. What is nauseating to us is proper to them.

Still more striking is our reaction to breaches of modesty. We have ourselves witnessed a marked change in regard to what is considered modest, what immodest. A comparative study shows that modesty is found the world over, but that the ideas of what is modest and what immodest vary incredibly. Thirty years ago woman's dress of to-day would have been immodest. South African Negroes greet a person of high rank by turning the back and bowing away from him. Some South American Indians consider it immodest to eat in view of other people. Whatever the form of modest behavior may be, a breach of etiquette is always strongly resented.

This is characteristic of all forms of automatic behavior. The performance of an automatic action is accompanied by the lowest degree of consciousness. To witness an action contrary to our automatic behavior excites at once intense attention and the strongest resistances must be overcome if we are required to perform such an action. Where motor habits are concerned the resistance is based on the difficulty of acquiring new habits, which is the greater the older we are, perhaps less on account of growing inadaptibility than for the reason that we are constantly required to act and have no time to adjust ourselves to new ways. In trifling matters the resistance may take the form of fear of ridicule, in more serious ones there may be dread of social ostracism. But it is not only the fear of the critical attitude of society that creates resistance, it rests equally in our own unwillingness to change, in our thorough disapprobation of the unconventional.

Intolerance of sharply divided social sets is often based on the strength of automatic reactions and upon the feeling of intense displeasure felt in acts opposed to our own automatism. The apparent fanaticism exhibited in the persecution of heretics must be explained in this manner. At a time when the dogma taught by the Church was imposed upon each individual so intensely that it became an automatic part of his thought and action, it was accompanied by a strong feeling of opposition, of hostility to any one who did not participate in this feeling. The term fanaticism does not quite correctly express the attitude of the Inquisition. Its psychological basis was rather the impossibility of changing a habit of thought that had become automatic and the consequent impossibility of following new lines of thought, which, for this very reason, seemed antisocial; that is, criminal.

We have a similar spectacle in the present conflict between nationalism and internationalism with their mutual intolerance.

Even in science a similar intolerance may be observed in the struggle of opposing theories and in the difficulty of breaking down traditional viewpoints.

Both the positive and negative effect of automatically established actions implies that a culture replete with these must be stable. Every individual behaves according to the setting of the culture in which he lives. When the uniformity of automatic reaction is broken, the stability of culture will be weakened or lost. Conformity and stability are inseparably connected. Non-conformity breaks the force of tradition.

We are thus led to an investigation of the conditions that make for conformity or non-conformity.

Conformity to instinctive activities is enforced by our organic structure, conformity to automatic actions by habit. The infant learns to speak by imitation. During the first few years of life the movements of larynx, tongue, roof of the mouth, and lips are gradually controlled and finally executed with great accuracy and rapidity. If the child is removed to a new environment in which another language is spoken, before the time when the movements of articulation have become stable, and as long as a certain effort in speech is still required, the movements required by the new language are acquired with perfect ease. For the adult a change from one language to another is much more difficult. The demands of everyday life compel him to use speech, and the articulating organs follow the automatic, fixed habits of his childhood. By imitation certain modifications occur, but a complete break with the early habits is extremely difficult, for many well-nigh impossible, and probably in no case quite perfect. Unwonted movements reappear when, due to disease, the control of the central nervous system breaks down.

Early habits control also the movements of the body. In childhood we acquire certain ways of handling our bodies. If these movements have become automatic it is almost impossible to change to another style, because all the muscles are attuned to act in a fixed way. To change one's gait, to acquire a new style of handwriting, to change the play of the muscles of the face in response to emotion is a task that can never be accomplished satisfactorily.

What is true of the handling of the body is equally true of mental processes. When we have learned to think in definite ways it is exceedingly difficult to break away and to follow new paths. For a person who has never been accustomed as a young child to restrain responses to emotions, such as weeping, or laughing, a transition to the restraints cultivated among us will be difficult. The teachings of earliest childhood remain for most people the dogma of adult life, the truth of which is never doubted. Recently the importance of the impressions of earliest childhood has been emphasized again by psychoanalysts. Whatever happens during the first five years of life sets the pace for the reactions of the individual. Habits established in this period become automatic and will resist strongly any pressure requiring change.

It would be saying too much to claim that these habits are alone responsible for the reactions of the individual. His bodily organization certainly plays a part. This appears most clearly in the case of pathological individuals or of those unusually gifted in one way or another; but the whole population consists of individuals varying greatly in bodily form and function, and since the same forms and faculties occur in many groups, the group behavior cannot be deeply influenced by structure. Differences must be due to culturally acquired automatic habits and these are among the most important sources of conservatism.

A few examples may illustrate the conditions that fix our habits. Fire-making by friction is known almost everywhere. Most people use the fire-drill which is turned backward and forward between the hands or by mechanical means. Others use the fire-plough, still others the fire-saw. The principle is always the same. Wood is rubbed against wood under pressure and with such rapidity that the dust produced by friction finally ignites. The motion applied is different; in one case drilling, in another ploughing or cutting, in a third sawing. Another example is the production of flour. Some people obtain it by grinding kernels between stones, others by pounding in a mortar. The forms of pestles for pounding depends upon material, the use of one or two hands, and upon the mode of holding the pestle. Hammers may be hafted or unhafted, used with one or two hands. Their purpose is always the same, but their forms differ according to the ways in which the hammer is customarily used. Some tribes use hand adzes with long handles, others those held close to the cutting blade. A draw knife is used for cutting towards the body, other forms of knives for whittling away from the body. For a person accustomed to cut with a draw knife, a knife handle not fitted for this movement is unhandy. The motor habits of people are reflected in the form of the handles of their tools.

The movements determined by the forms of handles are sometimes very special and a change to another form of handle is correspondingly difficult. A good example of this is the throwing board of the Eskimo. The board serves to give a greater impetus to a lance or a dart than the one that can be given by the hand. It is, as it were, an extension of the hand. The one end is held in the hand. On the surface is a groove in which the lance rests so that its butt end is supported at the other end. When the arm swings forward in the motion of throwing, the lance rests against the far end of the board, which, on account of its greater distance from the shoulder, moves more swiftly and thus gives greater impetus to the weapon. The accuracy with which the lance is thrown depends upon the intimate familiarity of the hand with the board, for the slightest variation in its position modifies the flight of the weapon. The forms of the throwing board differ considerably from tribe to tribe. In Labrador and in the region farther north it is broad and heavy, with grip holds for thumb and fingers. In Alaska it is slender with a grip arranged in quite a different manner. A hand accustomed to the wide board would require considerable time to learn the use of the narrower one. An implement of the same kind occurs in Australia, but its form is fundamentally different. I presume an Australian who would try to use an Eskimo throwing board would fail to hit his game.

The same is true of our modern tools. The movements of the body are adjusted to the handle of the tool. The handle was not changed until machinery was introduced. The handle of the plane looks as though it were adapted to the hand. Its form has developed so as to facilitate the movements which we use. If we should use a different kind of movement for planing the form of the handle would have to be different, too; but the use of the handle that has been developed fixes the habitual movements that we acquire.

Our posture may serve as another example. We sit on chairs. We like to have our backs supported and our feet on the floor. The Indians do not find this comfortable at all. They sit on the ground. Some stretch their legs forward, others sideways. Many squat down, bending the lower legs backward and sitting on the ground between the feet. For most adults, among ourselves, this position is impossible.

The form of furniture depends upon our habitual posture. Some people sleep on the back, others on the side. When sleeping on the side it is convenient to support the head with a pillow. People who sleep on the back find it convenient to support the neck by a narrow rest while the shoulders rest on the ground and the head is suspended. The neck rest cannot be used when it is customary to sleep on the side. The forms of chairs, beds, tables, and many kinds of household utensils are thus determined by our motor habits. They have developed as an expression of these habits, but their use compels every succeeding generation to follow the same habits. Thus they tend to stabilize them and to make them automatic.

The difficulty of changing forms dependent upon well-established motor habits is well illustrated by the permanence of the keyboard of the piano, which withstands all efforts at improvements; or by the complexity of forms and inadequacy of the number of symbols of our alphabet, which is hardly realized by most of those who write and read. In all these matters universality of habit in the social group brings about conformity of all the constituent individuals.

The most automatic activity of man is his speech and it is well worth while to inquire in how far habitual speech causes conformity of our actions and thought. The problem might also be so formulated that we ask in how far does language control action and thought, and in how far does our behavior control language. Some aspects of this question have been touched upon before (p. 139).

Language is so constituted that when new cultural needs arise it will supply the forms that express them. There is a large number of words in our vocabulary that have arisen with new inventions and new ideas that would be unintelligible to our ancestors who lived two hundred years ago. On the other hand, words no longer needed have disappeared.

What is true of words is equally true of forms. Many primitive languages are very definite in expressing ideas. Locality, time, and modality of any statement are denoted accurately. An Indian of Vancouver Island does not say "the man is dead," he would say "this man who has passed away lies dead on the floor of this house." He does not, according to the form of his language, express the idea "the man is dead" in generalized form. It might seem that this is a defect in his language, that he cannot form a generalized statement. As a matter of fact he has no need of generalized statements. He speaks to his fellow-men about the specific events of everyday life. He does not speak about abstract goodness, he speaks about the goodness of a certain person and he has no call to use the abstract term. The question is what happens when his culture changes and generalized terms are needed. The history of our own language shows clearly what does happen. We do not mind forcing the language into new molds and creating the forms that we require. If the philosopher develops a new idea he forces the language to yield devices that will adequately express his ideas and if these take root the language follows the lead thus given. A careful examination of primitive languages shows that these possibilities are always inherent in their structure. When missionaries train natives to translate the Bible and the Book of Prayer they compel them to do violence to the current forms; and it can always be done. In this sense we may say that culture determines language.

Most instructive in this respect are those parts of the vocabulary that express systems of classification; most notably in the numerical system and in the terminology of relationship.

All counting is based on a grouping of units. We group by tens and do so automatically. Some languages group by fives and combine four fives—that is the fingers and toes—in one higher unit. In English their terminology would be one, two, three, four, five; one, two, three, four, five on the other hand; one, two, three, four, five on the one foot; one, two, three, four, on the other foot; and finally, for our twenty, a man. If I want to say in such a language 973, I have to group the units not in 9 times 10 times 10 (900) plus 7 times 10 (70) plus 3, but in 2 times 20 times 20 (800) plus three on the other hand (=8) times 20 (160) plus three on the one foot (=13). In other words *we* count 973 units as 900+70+3. In the other language 973 are counted as 2×400 plus (5+3) times 20 plus (10+3). Every number is divided in groups of units, multiples of twenty, of 400, 8,000 and so on. To acquire this new classification automatically is an exceedingly difficult process.

Our terms of relationship are based on a few simple principles: generation, sex, direct descent or side line. My uncle is a person of the first ascendant generation, male, side line. Among other people the principles may be quite different. For instance, the difference between direct and side line may be disregarded, while the terms may differ according to the sex of the speaker. Thus a male calls his mother and all females of the first ascendant generation by one term, and also his sons and nephews by a single term. The concept and emotional significance of our term mother cannot persist in such a terminology. The adjustment to the new concepts that make impossible the customary automatic emotional reaction to terms of relationship will also be exceedingly difficult.

In another way language sways the forms of our thought. Every language has its own way of classifying sense experience and inner life, and thought is, to a certain extent, swayed by the associations between words. To us activities like breaking, tearing, folding may call forth the ideas of the kind of things that we break, tear or fold. In other languages the terms express with such vigor the way in which these actions are done, by pressure, by pulling, with the hand; or the stiffness, hardness, form, pliability of the object that the flow of ideas is determined in this fashion.

More important than this is the emotional tone of words. Particularly those words that are symbols of groups of ideas to which we automatically respond in definite ways have a fundamental value in shaping our behavior. They function as a release for habitual actions. In our modern civilizations the words patriotism, democracy or autocracy, liberty are of this class. The real content of many of these is not important; important is their emotional value. Liberty may be non-existent, the word-symbol will survive in all its power, although the actual condition may be one of subjection. The name democracy will induce people to accept autocracy as long as the symbol is kept intact. The vague concepts expressed by these words are sufficient to excite the strongest reactions that stabilize the cultural behavior of people, even when the inner form of culture undergoes considerable changes that go unnoticed on account of the preservation of the symbol.

Words are not the only symbols that influence behavior in this manner. There are also many objective symbols, such as the national flags or the cross, or fixed literary and musical forms that have attained the value of symbols, like the formal prayers of various creeds, national songs and anthems.

The conservative force of all of these rests on their emotional effect.

The uniformity of automatic reaction of the whole society is one of the strongest forces making for stability. When all react in the same way it becomes difficult for an individual to break away from the common habits. In a complex culture in which diverse attitudes are found the probability of change must be much greater.

This is strikingly illustrated by the contrast between the culture of primitive tribes and our modern civilization. Our society is not uniform. Among us even the best educated cannot participate in our whole civilization. Among primitive tribes the differences in occupations, interests, and knowledge are comparatively slight. Every individual is to a great extent familiar with all the thoughts, emotions, and activities of the community. The uniformity of behavior is similar to that expected among ourselves of a member of a social "set." A person who does not conform to the habits of thought and actions of his "set" loses standing and must leave. In our modern civilization he is likely to find another congenial "set" to the habits of which he can conform. In primitive society such sets are absent. With us the presence of many groups of different standards of interest and behavior is a stimulus for critical self-examination, for conflicts of group interests and other forms of intimate contact are ever present. Among primitive people this stimulus does not occur within the tribal unit. For these reasons individual independence is attained with much greater difficulty and tribal standards have much greater force.

Individual independence is the weaker the more markedly a culture is dominated by a single idea that controls the actions of every individual. We may illustrate this by the example of the Indians of the northwest coast of America and of those of the Plains. The former are dominated by the desire to obtain social prominence by the display of wealth and by occupying a position of high rank which depends upon ancestry and conformity to the social requirements of rank. The life of almost every individual is regulated by this thought. The desire for social prestige finds expression in amassing riches, in squandering accumulated wealth, in lavish display, in outdoing rivals of equal rank, in marrying so as to insure rank for one's children, more even than in a set of rich young people in our cities who have inherited wealth and who lose caste unless they come up to the social pace of their set. The uniformity of this background and the intensity with which it is cultivated in the young do not allow other forms to arise and keep the cultural outlook stable. Quite similar observations may be made among the natives of New Guinea, among whom display of wealth is also a dominating passion.

Quite different is the background of life of the Indians of the Plains. The desire to obtain honors by warlike deeds prompts thoughts and actions of everyone. Social position is intimately bound up with success in war, and the desire for prominence is inculcated in the mind of every child. The combination of these two tendencies determines the mental status of the community and prevents the development of different ideals.

Again different are conditions among the sedentary tribes of New Mexico. According to Dr. Ruth L. Bunzel the chief desire of the Zuñi Indian is to conform to the general level of behavior and not to be prominent. Prominence brings with it so many duties and enmities that it is avoided. The dominating interest in life is occupation with ceremonialism and this combined with fear of outstanding responsibility gives a steady tone to life.

The fundamental contrast between Pueblo formalism and the abandon to exaltation of other Indian tribes has been set forth clearly by Dr. Benedict. Among the Pueblos there is no desire to cultivate customs that lead to individual or mass excitement, no use of drugs to produce ecstasy, no orgiastic dance, no self-torture, no self-induced vision, traits that are common to almost all other Indian tribes.

No less instructive is the fundamental rôle played by the idea of the sacredness of persons of high rank, expressed particularly by the taboo of their persons and of objects belonging to them, that prevails practically all over Polynesia and that must be an ancient trait of Polynesian culture.

In all these cases the uniformity of social habits and the lack of examples of different types of behavior make deviations difficult and place in an antisocial class the individual who does not conform, even if his revolt is due to a superior mind and to strength of character.

In primitive society the general cultural outlook is in most cases uniform and examples that are opposed to the usual behavior are of rare occurrence. The participation of many in a uniform attitude has a stabilizing effect.

European history also shows conclusively that fundamental viewpoints once established are held tenaciously. Changes develop slowly and against strong resistance. The relation of the individual to the Church may serve as an example. The willing submission to Church authority which characterized European and American life in earlier times and the unhesitating acceptance of traditional dogma are giving way to individual independence, but the transition has been slow and is still vigorously resisted by those who adhere to the earlier attitude. The ease with which changes of denominational affiliation or complete break with the Church are accepted was unthinkable for many centuries and is even now resented by many.

The slow breaking up of feudalism and the gradual disappearance of the privileges of royalty and nobility are other pertinent examples.

The history of rationalism is equally instructive. The endeavor to understand all processes as the effects of known causes has led to the development of modern science and has gradually expanded over ever-widening fields. The rigid application of the method demands the reduction of every phenomenon to its cause. A purpose, a teleological viewpoint, and accident are excluded. It was probably one of the greatest attractions of the Darwinian theory of natural selection that it substituted for a purposive explanation of the origin of life forms a purely causal one.

The strength of the rationalistic viewpoint is also manifested in the attitude of psychoanalysis which refuses to accept any of our ordinary, everyday actions as accidental, but demands an inner, causal connection between all mental processes.

It would be an error to assume that the universal application of rationalism is the final form of thought, the ultimate result which our organism is destined to reach. Opposition to its negation of purpose, or its transformation of purpose into cause and to its disregard of accident as influencing the individual phenomenon, is struggling for recognition.

When at times of great popular excitement the masses in civilized society are swayed by a single idea, the independence of the individual is lost in the same manner as it is in primitive society. We have passed through a period of such dominant ideas during the World War and it is probable that every European nation was affected in the same manner. What seemed before the outbreak of hostilities as momentous differences vanished, and one thought animated every nation.

The compelling force of popular ideas is weaker in a diversified culture, in which the child is exposed to the influences of conflicting tendencies, so that none has the opportunity to become automatically settled, to become sufficiently firmly ingrained in nature to evoke intense resistance against different habits. When only one dominant attitude exists, the rise of a critical attitude requires a strong, creative mind. Where many exist and none has a marked, emotional appeal, opportunity for critical choice is given.

The greater the differentiation of groups within the social unit, and the closer the contact between them the less is it likely that any of the traditional lines of behavior will be so firmly established that they become entirely automatic. In a diversified culture the child, as long as it does not become a member of a sharply segregated set, is exposed to so many conflicting tendencies that few only have the opportunity to become so strongly ingrained in nature as to evoke energetic resistance against different habits. A stratified society consisting of loosely defined classes with privileges and different viewpoints is, therefore, more subject to change than a homogeneous society. When in a stratified society the sets are sharply segregated, so that they develop their own codes of behavior, their conservatism in regard to their specialized attitudes may easily equal that of unstratified societies, the more so the more exclusive they are. An example of this is the code of honor that prevailed until recently among the officers in European armies, which required the settlement by duel of disputes involving points of honor as understood by the class while judicial settlement was considered as dishonorable. Similar phenomena are not absent in primitive society. Thus the code of honor of the Crazy Dog Society among the Plains Indians required a type of bravery not expected from the ordinary warrior.

Lack of stratification may account for the intense conservatism of the Eskimo, whose culture has changed very little over a long period. They are remote from contact with foreign cultures, and their society is remarkably homogeneous, all households being practically on the same level and all participating fully in the tribal culture. In contrast to the permanence of their culture there is evidence of comparatively rapid changes among the Indians of British Columbia. They are exposed to contact with cultures of distinct types; and on account of the diversity of privileges of individuals, families and societies their customs have been in a state of flux.

Changes are facilitated in all those cases in which customs are entrusted to the care of a few individuals. Among many tribes sacred ceremonials are in the keeping of a few priests or of a single chief or priest. Although they are supposed to preserve the ceremonial faithfully in all its details, we have ample evidence showing that owing to forgetfulness, to ambition, to the workings of a philosophic or imaginative mind, or to the premature death of the keeper of the secret, the forms may undergo rapid changes.

The influence of an individual upon culture depends not only upon his strength but also upon the readiness of society to accept changes. During the unstable conditions of cultural life produced by contact between European and primitive civilizations opportunity is given to the individual to exert a marked influence upon tribal life. It is not easy to find instances in which a new invention may be attributed to a known individual, but evidence is available showing how suddenly a new element, suggested or invented by a native, or sometimes by an outsider, spreads. An invention of this kind is the lock designed by the Eskimos of Smith Sound, to replace broken gun locks which without the necessary tools and materials they were unable to repair. The change of form of their ivory harpoon head and harpoon shaft with ivory foreshaft, when iron came into use for both head and foreshaft, may have been made with the help of

American and Scotch whalers, but was quickly adopted.

Another example came to my notice in the winter 1930-1931. In former times all the Indians of Fort Rupert, British Columbia, lived in large, square plank houses. Some of these still exist. They are provided with small bedrooms arranged on a platform that runs inside around the walls. A large fireplace is in the center of the house. When a feast is held the central room is cleared and the people sit around the walls at the foot of the platform. Nowadays many of them live in frame houses so planned that a large front room, unfurnished, with a stove in the center serves as general assembly room. Kitchen, bedrooms, and storerooms are small and placed behind the front room. The plan was evidently designed to make possible the old type of assembly, for the guests sit around the walls on the floor in the same way as in the old plank houses. Among the same tribe a woman introduced about 1920 a new style of decorating the open spruce root burden baskets. The decoration is made by introducing broad splints of wood in the meshes. These are now being made by all the women who make basketry, and with the new decoration new forms of the baskets have developed. At the same time the weavers began to copy the imbricated basketry of the Lillooet, a tribe of the interior of British Columbia, and this also has been taken up widely.

Still more characteristic are the new forms of religious dogma and practice that have sprung up under modern conditions. Many native prophets have arisen who have, with greater or less success, modified the religious beliefs of the people. Their revelations, however, were reflexes of the mixed culture. Such was the ghost dance religion which originated among the Utes and spread over a large part of North America; such is the Shaker religion of the State of Washington, a church organized on Christian pattern, the dogma of which began as a curious blend of Christianity and ancient belief, but has more and more developed in the direction of the spirit of ancient shamanism. At present the spread of the peyote cult of the Indians exhibits the same characteristics. The visions induced by the drug, ancient beliefs, and Christian teachings have resulted in a variety of cults in which old and new are inextricably interwoven.

The influence of the individual upon art style may also be traced in a number of cases. Ordinarily the artist is hemmed in by the peculiar style of the art and technique of his environment. I had a number of Indian school children in a Government School at Alert Bay, British Columbia, draw figures of animals, without any suggestion regarding the animal to be chosen or the way it was to be drawn. Many of the boys twelve years old and older chose the killer whale and drew it according to the old style of Indian art which is so strongly impressed upon their minds that deviations are rare, although new combinations occur. The best known cases of new styles developed by individuals are those of Maria Martínez of the Pueblo of San Ildefonso who invented a new technique of dull design on a background of shining black combined with new patterns; and that of Nampeyo, a Hopi woman who created a new style of pottery based on the designs on shards of prehistoric pottery bowls.

A political leader may add new ideas to old political forms, although the older forms will exert an influence upon his mind and limit the extent to which the new may become acceptable. Thus the famous League of the Iroquois which in all probability was in its main outlines the creation of an individual, was based on the ancient social organization of the tribes. Perhaps the most outstanding example of this kind is the reorganization of the Zulu by Chaka who created a rigid military government.

Only when a new culture, a change of religion or of economic life is imposed by force, as was done by the Inca, or as happened in the early extension of Christianity and of Mohammedanism, and again in some regions during the forcible extermination of Protestantism, or as is happening now in Russia in its economic readjustment, may one group succeed in the attempt to impose radical changes in culture.

Ordinarily the new ideas created in a society are not free but directed by the culture in which they arise. Only when the culture is shaken by the impact of foreign ideas or by violent changes of culture owing to disturbing conditions is the opportunity given to the individual to establish new lines of thought that may give a new direction to cultural change.

It is hardly necessary to dwell again on the rapidity of recent changes in attitudes brought about by the advances in science and by the general spread of knowledge which favor rational critique of tradition and thus undermine many of the beliefs and customs that survive from earlier times. It is, however, worth remarking that, notwithstanding the decided breaking down of belief in tradition, strong resistances persist. These are well illustrated by the superstitious attitudes of college students, collected by Professor Tozzer, by the vogue of belief in plainly fraudulent spiritistic media, even among educated persons, and by the readiness of acceptance of Christian Science.

In general we may observe that actions are more stable than thoughts.

The ease with which words change their meanings while retaining their form which is produced by movements of the articulating organs is one of the many examples that may be adduced.

More striking examples are found in a variety of cultural facts. In North America similar rituals are performed over a wide area. The general plan and most of the details are the same among many tribes. They all do nearly the same things. On the other hand, the significance of the ritual differs considerably among various tribes. The so-called Sun Dance, which is alike in plan and the main features of its execution, serves in one tribe as a prayer for success in war; by another it is used as a pledge in prayers for recovery from serious illness. It is also a means of preventing disease.

The decorative art of the Plains Indians is another excellent example. The designs used in painting and embroidery are largely simple forms, such as straight lines, triangles, and rectangles. Their composition also is so much alike among many tribes that we must necessarily assume the same origin for the forms. We look at the designs as purely ornamental. To the Indian they have a meaning, somewhat in the same way as we associate a meaning with the flag and other national or religious emblems. The meanings, the thoughts connected with the designs are very variable. An isosceles triangle with short straight lines descending from its base suggests to one tribe a bear's paw with its long claws; to another a tent with the pegs that hold down the cover; to a third a mountain with springs at its foot; to a fourth a rain cloud with descending rain. The meaning changes according to the cultural interests of the people; the form which is dependent upon their industrial activities does not change.

The same observation may be made in the tales of primitive people. Identical tales are told over wide territories by people of fundamentally different types of culture. The ideas that attach themselves to a tale depend upon cultural interests. What is a sacred myth in one tribe is told for amusement in another. If the interest of the people centers in the stars we may have the tale as a star myth, if they are interested in animals it may explain conditions in the animal world; if they have at heart ceremonial life the tale will deal with ceremonies.

Secondary explanations are also common in our own civilization. We speak of some of the old customs that have lost or changed their meanings as "survivals." Many of the paraphernalia used by European royalty or by the Church are survivals of early times that have changed their meaning.

A good example is the history of food taboos. The Jewish taboo of animals other than ruminants with cloven hoofs is analogous to the food taboos of people the world over. Its origin is pre-Mosaic and has no relation to the development of Jewish monotheism. Still it is interpreted by Jewish orthodoxy as an important element of the Jahwe religion. In our rationalistic times the attempt is being made to explain the taboo of pork particularly on the basis of the alleged experience of its unwholesomeness in tropical countries.

Quite similar is the history of the taboo of incest. We do not know what its origin may be, but its breach is considered almost everywhere as one of the most heinous religious offences. Nowadays it is often naïvely assumed that it is based on the experience of the detrimental effect of inbreeding. This is certainly not its cause, for incest is not a biological but a sociological concept. It is not a question of preventing marriages between relations by blood, but rather between those who belong to groups considered as relatives. Thus marriages between the children of two brothers or those between the children of two sisters are often considered as incestuous, while among the same tribes the son of a woman is required to marry the daughter of her brother. Both groups are, biologically speaking, equally closely related.

An analogous change is developing in regard to Sunday. It is now considered a day of rest for people to recuperate from the work of the week. It originated as a holy day and is analogous to unlucky days or to days on which hostile tribes meet peacefully for the purpose of barter.

These customs must be considered as automatic, established by long-continued habits. When they are raised into consciousness our rationalizing impulses require a satisfying explanation and this follows the prevailing pattern of thought.

CHAPTER VIII

EDUCATION

WHEN investigating the physical characteristics of mankind, anthropologists do not confine themselves to the study of the adult. They investigate also the growth and development of the child. They record the increase in size of the body and of its organs, the changes in physiological reaction and of mental behavior. The results of these studies are laid down in certain norms characteristic of each age and each social or racial group.

Physiologically and psychologically the child does not function in the same way as the adult, the male not in the same way as the female. Anthropological research offers, therefore, a means of determining what may be expected of children of different ages and this knowledge is of considerable value for regulating educational methods. From this point of view Maria Montessori has developed a pedagogical anthropology and many educators occupy themselves with investigations of form and function of the body during childhood and adolescence, in the hope of developing standards by which we can regulate our demands upon the physiological and mental performances of the child. More than that, many educators hope to be enabled to place each individual child in its proper position and to predict the course of its development.

Anthropological investigations of an age class, let us say of eight-year-old children, show, for a selected social and racial group, a certain distribution of stature, weight, size of head, development of the skeleton, condition of teeth, size of internal organs and so on. The children represented in the group are not by any means equal, but each series of observations shows the majority of individuals ranging near a certain value and few exhibiting values of measurements remote from a middle value, the fewer the more remote from it. If the statures of eight-year-old boys range around forty-nine inches, then the number of those who are one, two, three inches taller or shorter than this value will be the fewer in number the greater excess or deficiency in stature. We have seen, before, in our consideration of races, that it is a mistake to consider the middle value as the norm. We must define the type by the distributions of the various measurements of the whole series of individuals included in our age class.

There are two causes that bring about variations in stature or other traits of growing children. The rate of growth is determined on the one hand by heredity; on the other hand it is strongly influenced by outer accelerating or retarding conditions, such as more or less adequate nutrition, the incidence of diseases and the amount of fresh air and sunshine enjoyed by the child.

When boys of different ages are compared—for instance, children of seven years and nine years of age with those of eight years of age of whom we spoke just now—it will be found that the range of forms in these three adjoining years is so wide that many sizes are found that belong to any one of the three age classes. This is true, not only of stature, but of all other measures, no matter whether we are dealing with anatomical or functional values. This merely expresses the common observation that the physical development of a child and its behavior do not allow us to give a correct estimate of its age.

The reasons for the differences between children are quite varied. Form and size of the body and its functioning depend upon heredity. Children of a tall family tend to be tall; children of a family of stocky build are liable to develop bodily form of the same type. The physical basis for similarity of function is also determined by heredity.

Another cause for differences is found in different environmental conditions. Food, sunshine, fresh air, accidental sickness or freedom from sickness are important contributory elements.

Differences in the rate of development may be due to hereditary constitution or to environmental conditions. These last are of particular importance in the application of anthropological standards to educational problems. If we could determine whether a child is retarded or accelerated in its development, and if we knew the standards for each age, the demands to be made upon the child could be regulated accordingly.

The rate of development of the individual is expressed primarily by the appearance of definite physiological changes. In a group of the same descent there is presumably a definite order in which physiological changes occur and deviations from this order may be interpreted as retardations or accelerations. We observe the ages at which certain changes in the body and in the functions of organs occur. The length of the period of gestation; the first appearance of teeth; the appearance of centers of ossification in the skeleton; the joining of separate bones, such as the shafts and ends of the long bones, fingers and toes; sexual maturity; the appearance of the wisdom teeth; are indications that, physiologically speaking, the respective parts of the body have reached a certain, definite state of development.

The time of occurrence of such phenomena has been studied to a certain extent, although not yet adequately. The observations show that at all ages the time at which these stages are reached, varies materially in different individuals, and the more so the later in life the particular stage develops. In fact, the degree of variation, even in childhood, is surprising. While the period of gestation varies only by days, the first appearance of the first teeth varies by many weeks. The time of the loss of the deciduous teeth differs by months and the period when maturity is reached differs by years. This variability of age at which definite physiological conditions are reached goes on increasing in later life. The signs of senility such as graying of the hair, climacterium, the flattening of the lense of the eye, the hardening of arteries, appear in different individuals many years apart. We speak, therefore, of a physiological age of an individual in contrast to his chronological age. If the normal age at which the permanent inner incisors of boys appear is seven and a half, then a six-year-old boy whose inner incisors are erupting is, physiologically speaking, seven and a half years old, or his physiological acceleration amounts to one and a half years, so far as tooth development is concerned.

If the whole body and its physiological and mental functions were developing as a unit we should have an excellent means of placing each individual according to his or her stage of development. Unfortunately this is not the case and an attempt to use a single trait for the determination of the physiological age of an individual will generally fail. Skeleton, teeth, and internal organs, while being influenced by the general state of development of the body, exhibit at the same time a considerable degree of independence which may be due to hereditary or to external causes.

The interrelation between the state of development of parts of the body is not known in detail. We do know that, in general, size and physiological age are related. Children who are adolescent are taller and heavier, in every respect larger, than children of the same age who do not yet show signs of approaching adolescence. The development of the skeleton is correlated with size, for among children of the same age the long bones of the taller ones approach mature stages more closely than those of the shorter ones. In a socially and racially homogeneous group the children whose permanent teeth erupt early are also taller on the average than those whose permanent teeth erupt late.

The same interrelation is expressed in the growth of children belonging to different social classes. The rapidity of the development of the body is closely related to the economic status of the family. The children of well-to-do parents, who enjoy plenty of food, exercise, fresh air and sunshine, develop more quickly than the children of the poor. Observations in Russia, Italy, America and in other countries all indicate that the time when a certain physiological stage is reached is earlier in the rich than in the poor. Therefore all the bodily measurements of children of the rich are greater than those of the poor of the same age and the differences between the two groups are greatest when growth is most rapid and the changes of physiological status are most pronounced. This happens during adolescence. Later on, when growth ceases, the rich are at a standstill, while the poor continue to grow, so that the difference between the groups is lessened, although it never disappears completely.

All this indicates that there is a correlation between the growth of different parts of the body. Still, these relations are subject to many disturbances. This has been observed particularly in regard to the teeth. The poor whose general development is retarded, shed their deciduous teeth earlier than the well-to-do—presumably on account of the greater care with which the deciduous teeth of children of the better situated classes are treated. Their deciduous teeth are carefully preserved, while those of the poor often decay and are lost. Therefore the stimulus for the early development of the permanent teeth due to the loss of the corresponding deciduous teeth does not occur among the well-to-do.

More important than the purely anatomical relations are those between the functions of the body and the state of bodily development. We have good evidence that these also are related. When we classify children of the same age according to their school standing, we find that those in the higher grades are much larger in every way than those in the lower grades. We also find that in regard to physiological status they are more advanced than children who are retarded in their school standing. Although this proof is not quite satisfactory, since the advancement in school will also depend upon the apparent bodily development of children, it indicates a rather interesting relation between the general functioning of the body and maturity.

A comparison between the two sexes from these

points of view shows that every physiological stage that has been investigated occurs earlier in girls than in boys. The difference in time is at first slight. The early stages of development of the skeleton observed during the first few years of life indicate a difference in favor of the girls of a few months. At the time of adolescence the physiological development of girls precedes that of boys by more than two years.

This difference is important. During the early years of childhood the apparent development of girls and boys, expressed by their stature and weight, is very nearly the same. From this observation the inference has been drawn that in early childhood the sex differences in size and form of the skeleton, muscles and so on are negligible, notwithstanding their importance in later life. If we compare, however, boys and girls at the same stage of physiological development, their relation appears quite differently. If a girl seven years old is at the same stage of physiological development as a boy eight years old, we should compare the bulk of the body at these stages, and not at the same chronological age. The boy of eight years is considerably taller and heavier than the girl of seven years. In other words, at the same stage of physiological development the relation of size characteristic of the sexes in adult life exists.

The correctness of this interpretation is proved by the measures of those parts of the body that grow slowly. Thus, on the average, the head of girls is always smaller than that of boys of the same age. In this case the actual ratio of the measures in the two sexes is not obscured because the increment of size corresponding to the amount of physiological acceleration of the girl is small as compared to the actual amount of sex difference; while in the case of weight and stature the corresponding increment is so great that it obscures the typical sex difference. The sex difference in the length of the head, measured from forehead to occiput, is about eight millimeters in favor of the men. The total increment due to growth for girls who may be in their physiological development two years ahead of boys is not more than about three millimeters. A sexual difference of five millimeters remains even during this period. The same relations appear in the slow-growing thickness of long bones which exhibit the same sex differences in childhood as in adult life.

These observations are important because they emphasize the existence in childhood of sexual differences in many parts of the body. These suggest the further question in how far the anatomical differences are accompanied by physiological and psychological differences.

What is true of physical measurements is equally true of mental observations: the powers of children increase rapidly with increasing age. The growing power of attention, of resistance to fatigue, the gradual increase of knowledge, the changes in form of thought, have been studied.

The practical value of all these investigations is

that they give us the means of laying out a standard of demands that may be made on boys and girls of various ages and belonging to a certain society. Particularly in an educational system of a large city the knowledge so gained is helpful in planning the general curriculum.

In a large educational system the observations on physiological age will also be helpful in assigning children a little more adequately to the grades into which they fit. It is probable that children of the same stage of physiological development will work together more advantageously than children of the same chronological age.

The existence of secondary sexual characteristics and the difference between the sexes in functional maturity should be considered in the problem of co-education. During the period of adolescence the physiological development of boys and girls of the same ages is so different that joint education seems of doubtful value. It would probably be of advantage to retain contact between boys and girls of equal maturity. The detailed plan of instruction should consider the differences between boys and girls.

We do not know much about differences in the rate of development determined by heredity, but it is not unlikely that these exist.

A comparison of some well-to-do Jewish children in New York and Northwest European children in Newark shows a slightly more rapid growth of the Jewish boys while they are young. With approaching adolescence the growth of Jewish boys slackens, while the Northwest Europeans continue to grow vigorously. The effect is that the statures of the adults are quite distinct. Among children of similar social groups maturity sets in at the same time among Jewish and non-Jewish children. There is no indication that the mode of life is essentially different. The same relation is found in a comparison of poor Hebrews and the mass of American public school children. Here also boys agree in their stature up to the fifteenth year. Then follows a period of rapid growth for the public school boys, and of retarded growth for the Jews.

Other differences have been observed in the growth of full blood Indians and half bloods. As children the former seem to be taller than the half bloods, while as adults the half bloods are taller than the Indians. It has also been shown that the increase of the size of the head differs in different racial groups. The data available at the present time are still very imperfect.

It is not by any means certain that these differences may not be due to environmental as much as to hereditary conditions. All we know with certainty is that when the adult forms of two races vary materially then the course of growth is also different.

It is probable that the characteristic periods when physiological changes occur may also differ among different races. The influence of outer conditions upon these phenomena is so great that nothing certain can be stated. The value of a knowledge of these phenomena for educational problems cannot be doubted.

Educators are not satisfied with the general result here outlined. They wish to ascertain the exact position of each individual in order to assign to him his proper place. This is more than the anthropological method can accomplish. Although a group of children may be segregated that are approximately at the same stage of physiological development, the individuals will not be uniform. This may be illustrated by a few examples.

Badly nourished children are on the whole smaller and lighter in weight than those well nourished. It is, therefore, likely that the small and light children of a certain age will include more undernourished individuals than the tall and heavy children. Undernourishment will also make children of a given age deficient in weight in comparison to their stature. It may then be expected that those who are small and light of weight in proportion to their size are more often undernourished than those showing the opposite traits.

According to this method, to which may be added a few other characteristics, undernourished children have been segregated and given better food to bring them up to the standard.

It is not difficult to prove that these criteria are not adequate and that errors may be expected. Children differ in bodily build by heredity. Some are tall with heavy bones, others small with a light skeleton.

These may be perfectly healthy and well nourished and still will appear in the "undernourished" class. Others may have been retarded in their early development by sickness and may be both too small and too light of weight. If we examine each individual carefully in regard to the appearance of skin and muscles and whatever indication can be found of undernourishment, we actually find a lack of agreement between the really undernourished group and the one segregated according to statistical methods. The group contains so many individuals who are tall and heavy that a tolerably accurate selection of the undernourished cannot be made by such means. Even if we consider the food that is given to each individual and include this criterion in our selection we do not succeed much better, because there are those who are well fed, but whose digestive system is at fault and who cannot make proper use of their food.

The selection will bring it about that a greater number of undernourished individuals are in the segregated class, but it would not be right to claim that in this manner all those who are undernourished have been found, nor that all those segregated are really undernourished. The individual investigation cannot be dispensed with.

The same conditions prevail in regard to all other characteristics. If the child is short of stature the shortness may depend upon hereditary smallness, upon retardation, or upon early unfavorable conditions which, however, may have been completely overcome.

Even when retardation can be proved by direct physiological evidence it does not follow that the child must belong mentally to the age class so indicated, for the conditions controlling physiological and psychological functioning are not by any means exclusively determined by physiological age. Hereditary character and environmental causes entirely independent of the time element are no less important. A group of children at exactly the same stage of physiological development as determined by the few available tests differ considerably among themselves. Their reactions may be quick or slow, their senses may be acute or dull, their experience may be so varied according to their home surroundings and general mode of life that a considerable variation in adaptability to educational requirements may be expected.

No matter what kind of measurements, experiments, and tests may be desired, their relation to the actual personality is always indirect. Without detailed study of the individual a proper pedagogical treatment is unattainable.

What is true of a group cannot be applied to an individual.

It will be seen that this agrees with our judgment regarding the significance of racial characteristics. We are apt to consider as characteristic of the group those features or measurements around which the great mass of individuals cluster. We believe that this is the type to which all conform. In doing so we forget that a wide range of variations is characteristic of every group and that a considerable number of individuals deviate widely from the "type," and that nevertheless these belong to the same group. For this reason the group standard cannot be applied to every individual. If, for practical reasons, as in education, it is desired to form a homogeneous group, the component individuals must be selected among different groups according to the characteristics that seem of importance.

There are cases in which for the sake of efficiency anthropological grouping may be utilized. When it is necessary to select large numbers from a population, as, for instance, for enlistment during the late war, it is useful to know that individuals of an unfavorable body build are on the whole not able to withstand the strain of army life. Very tall, slim persons with a slight depth of chest are of this kind. The flatter the chest the more of them will be unable to fulfill the demands made on bodily strength and endurance. It will then be economical to discard the whole class rather than to take advantage of the few who may be useful.

Similar considerations are valid in the selection of laborers for those employers who rate the laborer not as a person but solely according to his money value, because the turnover of labor will be less rapid if the adaptable individuals are numerous in the class from which the selection is made.

Educators are interested in another problem. It is desirable to predict the development of an individual. If a child has difficulties in learning, will it continue to be a dullard or may a better prognosis be given; or if a child is underdeveloped will it continue to remain puny?

The answer can be given at least to the physical side of this question. We have followed a considerable number of children from early growth on. A group of small young children are liable to grow less than tall children of the same age. During adolescence a group of tall children will grow less than a group of short children of the same age. The latter condition expresses clearly that the short children are on the whole physiologically younger than the tall ones and are, therefore, still growing while the taller ones are nearly mature. It can also be shown that children of a certain stature at a given age, who are accelerated in their growth, belong by heredity to a shorter type than those of the same group who are retarded. For a whole group it is possible to predict the average rate of growth, if the size at a given time and the amount of acceleration or retardation are known. However, these results are not significant for the individual. The causes by which the whole course of growth is controlled are too varied, the accidents that influence it cannot be predicted. It is true that the course of undisturbed development depends upon the hereditary character of the individual, but the varying environmental conditions disturb this picture.

What is true of the growth of the body is much more true of its functions, particularly of the mental functioning. A prediction of the future development of a normal individual cannot be made with any degree of assurance.

Anthropology throws light upon an entirely different problem of education. We have discussed before the causes that make for cultural stability and found that automatic actions based on the habits of early childhood are most stable. The firmer the habits that are instilled into the child the less they are subject to reasoning, the stronger is their emotional appeal. If we wish to educate children to unreasoned mass action, we must cultivate set habits of action and thought. If we wish to educate them to intellectual and emotional freedom care must be taken that no unreasoned action takes such habitual hold upon them that a serious struggle is involved in the attempt to cast it off.

The customary forms of thought of primitive tribes show us clearly how an individual who is hemmed in on all sides by automatic reactions may believe himself to be free. The Eskimo present an excellent example of these conditions. In their social life they are exceedingly individualistic. The social group has so little cohesion that we have hardly the right to speak of tribes. A number of families come together and live in the same village, but there is nothing to prevent any one of them from living and settling at another place with other families of his acquaintance. In fact, during a period of a lifetime the families constituting an Eskimo village are shifting about; and while after many years they generally return to the places where their relatives live, the family may have belonged to a great many different communities. There is no authority vested in any individual, no chieftaincy, and no method by which orders, if they were given, could be enforced. In short, so far as human relations are concerned, we have a condition of almost absolute anarchy. We might, therefore, say that every single person within the limits of his own mental ability and physical competency is entirely free to determine his own mode of life and his own mode of thinking.

Nevertheless it is easily seen that there are innumerable restrictions determining his behavior. The Eskimo boy learns how to handle the knife, how to use bow and arrow, how to hunt, how to build a house; the girl learns how to sew and mend clothing and how to cook; and during all their lives they apply the methods learned in childhood. New inventions are rare and the whole industrial life of the people runs in traditional channels.

What is true of their industrial activities is no less true of their thoughts. Certain religious ideas have been transmitted to them, notions of right and wrong, amusements and enjoyment of certain types of art. Any deviation from these is not likely to occur. At the same time, and since all alien forms of behavior are unknown to them, it never enters into their minds that any different way of thinking and acting would be possible, and they consider themselves as perfectly free in regard to all their actions.

Based on our wider and different experience we know that the industrial problems of the Eskimo might be solved in a great many other ways and that their religious traditions and social customs might be quite different from what they are. From the outside, objective point of view, we see clearly the restrictions that bind the individual who considers himself free.

It is not difficult to see that the same conditions prevail among ourselves. Families and schools which assiduously cultivate the tenets of a religious faith and of a religious ceremonial and surround them with an emotional halo raise, on the whole, a generation that follows the same path. The Catholicism of Italy, the Protestantism of Scandinavia and Germany, the Mohammedanism of Turkey, the orthodox Judaism, are intelligible only on the basis of a lack of freedom of thought due to the strength of the automatic reaction to impressions received in early childhood that exclude all new viewpoints. In the majority of individuals who grow up under these conditions a new, distinct viewpoint is not brought out with sufficient vigor to make it clear that theirs is not freely chosen, but imposed upon them; and, *if* strange ideas are presented, the emotional appeal of the thoughts that are part of their nature is sufficient to make any rationalization of the habitual attitude acceptable, except to those of strong intellect and character. To say the least, the cultivation of formal religious attitudes in family and school makes difficult religious freedom.

What is true of religion is equally true of subservience to any other type of social behavior. Only to a limited extent can the distribution of political parties be understood by economic considerations. Often party affiliation is bred in the young in the same way as denominational allegiance. This is certainly true in many parts of America. It is equally true among a large part of the former privileged classes of Europe and among part of the European peasantry. In other cases peculiar novel combinations of old ideas and new tendencies based on changed social or economic conditions arise, such as nationally or denominationally conservative and socially radical parties. Without the strength of the traditional nationalistic or religious background these can hardly be understood.

With the weakening of the energy with which definite ideas are impressed upon the young and familiarity with many varying forms develops the freedom of choice. The weakening of the valuation of the dogma and the spread of scientific information has resulted in the loss of cohesion of the Protestant churches.

The methods of education chosen depend upon our ideals. The imperialistic State that strives for power and mass action wants citizens who are one in thought, one in being, swayed by the same symbols. Democracy demands individual freedom of the fetters of social symbols. Our public schools are hardly conscious of the conflict of these ideas. They instill automatic reactions to symbols by means of patriotic ceremonial, in many cases by indirect religious appeal and too often through reaction to the example of the teacher that is imitated. At the same time they are supposed to develop mind and character of the individual child. No wonder that they fail in the one or the other direction, generally in the education to freedom of thought, or that they create conflicts in the minds of the young, conflicts between the automatic attitudes that are carefully nursed and the teachings that are to contribute to individual freedom.

It may well be questioned whether the crises that are so characteristic of adolescent life in our civilization and that educators assume to be organically determined, are not due in part to these conflicts, in part to the artificial sexual restraints demanded by our society. We are altogether too readily inclined to ascribe to physiological causes those difficulties that are brought about by cultural interference with the physiological demands of the body. It is necessary that the crises and struggles that are characteristic of individual life in our society be investigated in societies in which our restraints do not exist while others may be present, before we assume all too readily that these are inherent in "human nature."

The serious mental struggle induced by the conflict between instinctive reaction and traditional social ethics is illustrated by a case of suicide among the Eskimo. A family had lost a child in the fall and according to custom the old fur clothing had to be thrown away. Skins were scarce that year and a second death in the family would have led to disaster to all its members. This induced the old, feeble grandmother, a woman whom I knew well, to wander away one night and to expose herself, in a rock niche, to death by freezing, away from her children and grandchildren, to avoid their contamination by contact with a corpse. However, she was missed, found and brought back. She escaped a second time and died before she was found.

Another case is presented by the Chuckchee of Siberia. They believe that every person will live in future life in the same condition in which he finds himself at the time of death. As a consequence an old man who begins to be decrepit wishes to die, so as to avoid life as a cripple in the endless future; and it becomes the duty of his son to kill him. The son believes in the righteousness of his father's request. At the same time, he feels the filial love for his father—perhaps better, to a respected member of the small community to which he himself belongs—and a conflict arises between dutiful love and the traditional customs of the tribe. Generally the customary behavior is obeyed, but not without severe struggles.

An instructive example of the absence of our difficulties in the life of adolescents and the occurrence of others is found in the studies of Dr. Margaret Mead on the adolescents of Samoa. With the freedom of sexual life, the absence of a large number of conflicting ideals, and an easy-going attitude towards life, the adolescent crisis disappears, while new difficulties originate at a later period when complexities of married life develop. A similar example is presented in the life of one of our southwestern Indian tribes, the Zuñi, among whom, according to Dr. Ruth L. Bunzel, the suppression of ambition, the desire to be like one's neighbor and to avoid all prominence are cultivated. They lead to a peculiar impersonal attitude and to such an extent of formalism that individual crises are all but suppressed.

In a stable society we do not often find that conflict between generations which has been lamented for centuries by the old who praise the ideals and customs of their youth. Apparently this conflict is more acute now than in former times. If this is true it is probably due to the greater rapidity of cultural change of our times. It is particularly pronounced when the parents are brought up in a culture radically different from the one in which their children grow up. In America this happens with great frequency among immigrants raised in conservative, rural parts of Europe, while their children grow up in American cities and are educated in American schools. In stable, homogeneous culture youthful licentiousness may sometimes lead to conflicts of a different character between old and young.

We do not know enough about these questions, but our anthropological knowledge justifies the most serious doubts regarding the physiological determination or the necessity of occurrence of many of the crises and struggles that characterize individual life in our civilization. A thorough study of analogous situations in foreign cultures will do much to clear up this problem which is of fundamental importance for the theory of education.

It is a question whether the doubts that beset the individual in such a period are beneficial or a hindrance. The seriousness of the struggle is certainly undesirable and an easier transition will be facilitated by lessening the intensity of attachment to the situation against which he is led to rebel.

The lack of freedom in our behavior is not confined to the uneducated, it prevails in the thoughts and actions of all classes of society.

When we attempt to form our opinions in an intelligent manner, we are inclined to accept the judgment of those who by their education and occupation are compelled to deal with the questions at issue. We assume that their views must be rational and based on an intelligent understanding of the problems. The foundation of this belief is the tacit assumption that they have special knowledge and that they are free to form perfectly rational opinions. However, it is easy to see that there is no social group in existence in which such freedom prevails.

The behavior in somewhat complex primitive societies in which there is a distinction between different social classes, throws an interesting light upon these conditions. An instance is presented by the Indians of British Columbia, among whom a sharp distinction is made between people of noble birth and common people. In this case the traditional behavior of the two classes shows considerable differences. The social tradition that regulates the life of the nobility is somewhat analogous to the social tradition in our society. A great deal of stress is laid upon strict observance of convention and upon display, and nobody can maintain his position in high society without an adequate amount of ostentation and without strict regard for conventional conduct. These requirements are so fundamental that an overbearing conceit and a contempt for the common people become social requirements of an important chief. The contrast between the social proprieties for the nobility and those for the common people is very striking. Of the common people are expected humbleness, mercy and all those qualities that we consider amiable and humane.

Similar observations may be made in all those cases in which, by a complex tradition, a social class is set off from the mass of the people. The chiefs of the Polynesian Islands, the kings of Africa, the medicine men of many countries, present examples in which the line of conduct and thought of a social group is strongly modified by their segregation from the mass of the people. They form closed societies. On the whole, in societies of this type, the mass of the people consider as their ideal those actions which we should characterize as humane; not by any means that all their actions conform to humane conduct, but their valuation of men shows that the fundamental altruistic principles which we recognize are recognized by them too. Not so with the privileged classes. In place of the general humane interest the class interest predominates; and while it cannot be claimed that their conduct, individually, is selfish, it is always so shaped that the interest of the class to which a person belongs prevails over the interest of society as a whole. If it is necessary to secure rank and to enhance the standing of the family by killing off a number of enemies, there is no hesitation felt in taking life. If the standards of the class require that its members should not perform menial occupations, but should devote themselves to art or learning, then all the members of the class will vie with one another in the attainment of these achievements. It is for this reason that every segregated class is much more strongly influenced by special traditional ideas than is the rest of the people; not that the multitude is free to think rationally and that its behavior is not determined by tradition; but the tradition is not so specific, not so strictly determined in its range, as in the case of the segregated classes.

I believe this observation is of great importance when we try to understand conditions in our own society. Its bearing upon the problem of the psychological significance of nationalism will at once be apparent; for the nation is also a segregated class, a closed society, albeit segregated according to other principles; and the characteristic feature of nationalism is that its social standards are considered as more fundamental than those that are general and human, or rather that the members of each nation like to assume that their ideals are or should be the true ideals of mankind. The late President Wilson once gave expression to this misconception when he said that, if we—Americans—hold ideals for ourselves, we should also hold them for others, referring in that case particularly to Mexico. At the same time it illustrates clearly that we should make a fundamental mistake if we should confound class selfishness and individual selfishness; for we find the most splendid examples of unselfish devotion to the interests of the nation, heroism that has been rightly praised for thousands of years as the highest virtue, and it is difficult to realize that nevertheless the whole history of mankind points in the direction of a *human* ideal as opposed to a *national* ideal. And indeed may we not continue to admire the self-sacrifice of a great mind, even if we transcend to ideals that were not his, and that perhaps, owing to the time and place in which he lived, could not be his?

Our observation has also another important application. The industrial and economic development of modern times has brought about a differentiation within our population that has never been equalled in any primitive society. The occupations of the various parts of a modern European or American population differ enormously; so much so that in many cases it is almost impossible for people speaking the same language to understand one another when they talk about their daily work. The ideas with which the scientist, the artist, the tradesman, the business man, the laborer operate are so distinctive that they have only a few fundamental elements in common. The mathematician, chemist, biologist, physician, and engineer are understood only by fellow students. Ordinarily they do not understand the terminology of the banker, accountant, tailor, farmer, hunter, fisherman, or cook, unless their occupations happen to make them acquainted with one or the other of these trades and occupations. Here it may again be observed that those occupations which are intellectually or emotionally most highly specialized require the longest training, and training always means an infusion of historically transmitted ideas. Even in their own disciplines the majority are strongly influenced by traditional teaching. Evidences of this are the rise and decline of schools of thought and fashions in lines of research. More important is the effect of specialization. Critical study of one branch of science does not seem to engender a critical attitude in regard to other aspects of culture. It would seem that in altogether too many minds the critical faculty remains confined to a very narrow range and that outside of it faith in tradition and emotional yielding to popular views reigns supreme. It is therefore not surprising that the thought of what we call the educated classes is controlled essentially by those ideals which have been transmitted to us by past generations. These ideals are always highly specialized, and include the ethical tendencies, the esthetic inclinations, the intellectuality, and the expression of volition of past times. After long continued education according to these standards their control may find expression in a dominant tone which determines the whole mode of thought and which, for the very reason that it has come to be ingrained in our whole mentality, never rises into our consciousness.

In those cases in which our reaction is more conscious, it is either positive or negative. Our thoughts may be based on a high valuation of the past, or they may be in revolt against it.

When we bear this in mind we may understand the characteristics of the behavior of the intellectuals. It is a mistake to assume that their mentality is, on the average, appreciably higher than that of the rest of the people. Perhaps a greater number of independent minds find their way into this group than into some other group of individuals who are moderately well-to-do; but their average mentality is surely in no way superior to that of the workingmen who by the conditions of their youth have been compelled to subsist on the produce of their manual labor. In both groups mediocrity prevails; unusually strong and unusually weak individuals are the exceptions. For this reason the strength of character and intellect that is required for vigorous thought on matters in which intense sentiments are involved is not commonly found—among the intellectuals or in any other part of the population. This condition, combined with the thoroughness with which the intellectuals have imbibed the traditions of the past, makes the majority of them in all nations conventional. It has the effect that their thoughts are based on tradition, and that the range of their vision is liable to be limited.

There are of course strong minds among the intellectuals who rise above the conventionalism of their class, and attain that freedom that is the reward of a courageous search for truth, along whatever path it may lead.

In contrast to the intellectuals, the masses in our modern city populations are less subject to the influence of traditional teaching. Many children are so irregular in their school attendance, so little interested in their school work, or torn away from school so soon that it cannot make an indelible impression upon their minds, and they may never have known the strength of the conservative influence of a home in which parents and children live a common life. The more heterogeneous the society in which they live, and the more the constituent groups are free from historic influences; or the more they represent different historic traditions, the less strongly will they be attached to the past.

This does not preclude the possibility of the formation of small, self-centered, closed societies, among the uneducated, such as local isolated communities, or gangs that equal primitive man in the intensity of their group feeling and in the disregard of the rights of the outsider. On account of their segregation they no longer belong to the masses.

It would be an exaggeration if we should extend the view just expressed over all aspects of human life. I am speaking here only of those fundamental concepts of right and wrong that develop in the segregated classes and in the masses. In a society in which beliefs are transmitted with great intensity the impossibility of treating calmly the views and actions of the heretic is shared by both groups. When, through the progress of scientific thought, the foundations of dogmatic belief are shaken among the intellectuals and not among the masses, we find the conditions reversed and greater freedom of traditional forms of thought among the intellectuals—at least in so far as the current dogma is involved. It would also be an exaggeration to claim that the masses can sense the right way of attaining the realization of their ideals, for these must be found by painful experience and by the application of knowledge. However, neither of these restrictions touches our main contention; namely, that the desires of the masses are in a wider sense human than those of the classes.

It is therefore not surprising that the masses of a city population, whose attachment to the past is comparatively slight, respond more quickly and more energetically to the urgent demands of the hour than the educated classes, and that the ethical ideals of the best among them are human ideals, not those of a segregated class. For this reason I should always be more inclined to accept, in regard to fundamental human problems, the judgment of the masses rather than the judgment of the intellectuals, which is much more certain to be warped by unconscious control of traditional ideas. I do not mean to say that the judgment of the masses would be acceptable in regard to every problem of human life, because there are many which, by their technical nature, are beyond their understanding; nor do I believe that the details of the right solution of a problem can always be found by the masses; but I feel strongly that the problem itself, as felt by them, and the ideal that they want to see realized, is a safer guide for our conduct than the ideal of the intellectual group that stand under the ban of an historical tradition that dulls their feeling for the needs of the day.

One danger lurks in the universality of these reactions to human needs. The economic conditions in the civilized world are so much the same that, without attachment to an individualized, historically founded culture a uniformity of cultural desires and levels may be reached that would deprive us of the valuable stimulus resulting from the interaction of distinctive cultural forms. Already the lack of individuality of cities of moderate size weighs heavily on our lives. The fulfilment of elementary desires that are much the same the world over must find their counterpoise in the development of individuality in form and content.

One word more, in regard to what might be a fatal misunderstanding of my meaning. If I decry unthinking obedience to the ideals of our forefathers, I am far from believing that it will ever be possible or that it will even be desirable, to cast away the past and to begin anew on a purely intellectual basis. Those who think that this can be accomplished do not, I believe, understand human nature aright. Our very wishes for changes are based on criticism of the past, and would take another direction if the conditions under which we live were of a different nature. We are building up our new ideals by utilizing the work of our ancestors, even where we condemn it, and so it will be in the future. Whatever our generation may achieve will attain in course of time that venerable aspect that will lay in chains the minds of our successors, and it will require new efforts to free a future generation of the shackles that we are forging. When we once recognize this process, we must see that it is our task not only to free ourselves of traditional prejudice, but also to search in the heritage of the past for what is useful and right, and to endeavor to free the mind of future generations so that they may not cling to our mistakes, but may be ready to correct them.

CHAPTER IX

MODERN LIFE AND PRIMITIVE CULTURE

IN THE preceding pages we have considered the effect of a number of fundamental biological, psychological, and social factors upon modern problems.

There are many other aspects of modern culture that may be examined from an anthropological point of view.

One of the great difficulties of modern life is presented by the conflict of ideals; individualism against socialization; nationalism against internationalism; enjoyment of life against efficiency; rationalism against a sound emotionalism; tradition against the logic of facts.

We may discern tendencies of change in all these directions; and changes that appear to one as progress appear to another as retrogression. Attempts to further individualism, to restrict efficiency, to make tradition more binding would be considered as objectionable and energetically resisted by many. What is desirable depends upon valuations that are not universally accepted.

Such differences of opinion do not exist in the

domain of physics or chemistry. The purposes to which we apply physical or chemical knowledge are definite. We have certain needs that are to be filled. A bridge is to be built, houses are to be constructed, machinery for accomplishing some specific work is required, communication is to be facilitated, dyes are to be made, fertilizers to be invented. Some inventions create new needs that crave to be satisfied by further inventions. Always a definite object is to be attained, the value of which lies in the improvement of the outer conditions of life. As long as we are satisfied that the resulting comforts and facilities are desirable, the application of our knowledge is valuable. The importance of achievements based on advances in physical sciences is readily acknowledged in so far as they enable us to overcome obstacles that would beset our lives if we had to do without them.

The applicability of the results of research to practical problems of social life are similar when we consider aims universally recognized as desirable. Individual health depending upon the health of the whole group is perhaps one of the simplest of these. Even in this case difficulties arise. There are individuals of impaired health whose existence may somewhat endanger public health. Is it of greater value to segregate these from the social body to their disadvantage, or to run the slight risk of their unfavorable influence upon the whole population? The answer to this question will depend upon valuations that have no basis in science, but in ideals of social behavior, and these are not the same for all members of a modern social group.

In general we may say that in the practical application of social science absolute standards are lacking. It is of no use to say that we want to attain the greatest good for the greatest number, if we are not able to come to an agreement as to what constitutes the greatest good.

This difficulty is strongly emphasized as soon as we look beyond the confines of our own modern civilization. The social ideals of the Central African Negroes, of the Australians, Eskimo, and Chinese are so different from our own that the valuations given by them to human behavior are not comparable. What is considered good by one is considered bad by another.

It would be an error to assume that our own social habits do not enter into judgments of the mode of life and thought of alien people. A single phenomenon like our reaction to what we call "good manners" illustrates how strongly we are influenced by customary behavior. We are exceedingly sensitive to differences in manners; definite table manners, etiquette of dress, a certain reserve, are peculiar to us. When different table manners, odd types of dress, and an unusual expansiveness are found, we feel a revulsion and the valuation of our own manners tinges our description of the alien forms.

The scientific study of generalized social forms requires, therefore, that the investigator free himself from all valuations based on our culture. An objective, strictly scientific inquiry can be made only if we succeed in entering into each culture on its own basis, if we elaborate the ideals of each people and include in our general objective study cultural values as found among different branches of mankind.

Even in the domain of science the favorite method of approaching problems exerts a dominating influence over our minds. This is well illustrated by the fashions prevailing in different periods: the dialectics of the Middle Ages were as satisfying to the average scientific minds of that period as is the aversion to dialectics and the insistence on observation in modern times. The concentration of biological thought upon problems of evolution in the early Darwinian period presents another example. The kaleidoscopic changes in interest, foremost in physiological and psychological inquiries of our times,—such as the theories based on the functions of glands of internal secretion, on racial and individual constitution, or on psychoanalysis,—are others. The passionate intensity with which these ideas are taken up, leading to a temporary submersion of all others and to a belief in their value as a sufficient basis of inquiry, proves how easily the human mind is led to the belief in an absolute value of those ideas that are expressed in the surrounding culture.

The reasons for this type of behavior are not far to seek. We are apt to follow the habitual activities of our fellows without a careful examination of the fundamental ideas from which their actions spring. Conformity in action has for its sequel conformity in thought. The emancipation from current thought is for most of us as difficult in science as it is in everyday life.

The emancipation from our own culture, demanded of the anthropologist, is not easily attained, because we are only too apt to consider the behavior in which we are bred as natural for all mankind, as one that must necessarily develop everywhere. It is, therefore, one of the fundamental aims of scientific anthropology to learn which traits of behavior, if any, are organically determined and are, therefore, the common property of mankind, and which are due to the culture in which we live.

We are taught to lay stress upon national differences that occur among Europeans and their descendants. Notwithstanding the peculiarities characteristic of each nation or local division the essential cultural background is the same for all of these. The cultural forms of Europe are determined by what happened in antiquity in the Eastern Mediterranean. In our modern civilization we have to recognize the progeny of Greek and Roman culture. The slight local variations are built up on a fundamental likeness. They are insignificant when we compare them with the differences that obtain between Europe and peoples that have not developed on the basis of the ancient Mediterranean culture. Even India and China cannot be entirely separated from the historical influences emanating from western Asia and the Mediterranean area.

The objective study of types of culture that have developed on historically independent lines or that have grown to be fundamentally distinct enables the anthropologist to differentiate clearly between those phases of life that are valid for all mankind and others that are culturally determined. Supplied with this knowledge he reaches a standpoint that enables him to view our own civilization critically, and to enter into a comparative study of values with a mind relatively uninfluenced by the emotions elicited by the automatically regulated behavior in which he participates as a member of our society.

The freedom of judgment thus obtained depends upon a clear recognition of what is organically and what culturally determined. The inquiry into this problem is hampered at every step by our own subjection to cultural standards that are misconstrued as generally valid human standards. The end can be reached only by patient inquiry in which our own emotional valuations and attitudes are conscientiously held in the background. The psychological and social data valid for all mankind that are so obtained are basal for all culture and not subject to varying valuation.

The values of our social ideals will thus gain in clarity by a rigid, objective study of foreign cultures.

If we could be sure that, after the organically determined behavior has been discovered, the study of distinct cultural forms would ultimately lead to the discovery of definite laws governing the historical development of social life, we might hope to construe a system for a reasonable treatment of our social problems. It is, however, questionable whether such an ideal is within our reach.

The fundamental difficulty may be illustrated by examples taken from the inorganic world. When we express a law in physics or chemistry we mean that, certain conditions being given, a definite result will follow. I release an object at a given place and it will fall with definite speed and acceleration. I bring two elements into contact and they will form, under controlled conditions, a definite compound. The result of an experiment may be predicted if the conditions controlling it are known. If our knowledge of mechanics and mathematics is sufficient and the position of all the planets at one given moment is known, we can foretell what movements are going to happen and what movements happened in the past, as long as no disturbing outer influences make themselves felt.

Social phenomena cannot be subjected to experiment. Controlled conditions, excluding disturbing outer influences, are unattainable. These complicate every process that we try to study.

The more complex the phenomena the more difficult it is to foretell the future from a condition existing at a given moment, even if the essential laws governing the happenings are known. Supposing, for instance, we are studying erosion on a mountainside. Can we foretell which course it is to take, or how the present forms have resulted? We find a gulch. At its head is a large boulder that deflected the water and caused it to cut a channel for itself on one side. If the stone had not been there, the gulch would have had a different direction. It so happens that the soil in one direction was soft so that the running water cut readily into it. We are dealing solely with the laws of erosion, but even the most intimate knowledge of these cannot adequately explain the present course of the gulch. The boulder may be in its place because it was loosened by an animal walking along the mountainside. It fell down and rested at the place where it obstructed the course of the running water.

All incidents of this class that influence the isolated process we want to study are excluded in experimentation. They are accidents in so far as they have no logical relation to the process about which we desire to gain knowledge. Even in the astronomical problem just alluded to the positions of the heavenly bodies at the initial moment are in this sense accidental, because they cannot be derived from any mechanical law. Disturbing outer influences that have no relation to the law must be admitted as accidents that determine the distribution of matter at the moment chosen as the initial one.

These conditions make prediction of what is going to happen in a special case exceedingly difficult, if not impossible. Accidental occurrences that are logically not related to the phenomena studied modify the sequence of events that might be determined if the conditions were absolutely controlled and protected against all outside interference. This condition is attained in a completely defined physical or chemical experiment, but never in any phenomenon of nature that can only be observed, not controlled. Notwithstanding the advances in our knowledge of the mechanics of air currents, weather prediction remains uncertain in regard to the actual state of the weather at a given hour in a given spot. A general, fairly correct prognosis for a larger area may be possible, but an exact sequence of individual events cannot be given. Accidental causes are too numerous to allow of an accurate prediction.

What is true in these cases is ever so much more true of social phenomena. Let us assume that there exists a society that has developed its culture according to certain laws discovered by a close scrutiny of the behavior of diverse societies. For some reason, perhaps on account of hostile attacks that have nothing to do with the inner workings of the society, the people have to leave their home and migrate from a fertile country into a desert. They have to adjust themselves to new forms of life; new ideas will develop in the new surroundings. The fact that they have been transplanted from one region to another is just an accident—like the loosened boulder that determined the direction of the gulch.

Even a hasty consideration of the history of man

shows that accidents of this kind are the rule in every society, for no society is isolated but exists in more or less intimate relations to its neighbors.

The controlling conditions may also be of quite a different nature. The game on which the people subsist may change its habitat or become extinct, a wooded area may become open country. All cases of change of geographical or economic environment entail changes in the structure of society, but these are accidental events in no way related to the inner working of the society itself.

As an example we may consider the history of Scandinavia. If we try to understand what the people are at the present time we have to inquire into their descent. We must consider the climatic and geographic changes that have occurred since the period when the glaciers of the pleistocene retracted and allowed man to settle, the changes in vegetation, the early contact with southern and eastern neighbors. All these have no relation to the laws that may govern the inner life of society. They are accidents. If the Central Europeans had had no influence whatever upon Scandinavia the people would not be what they are. These elements cannot be eliminated.

For these reasons every culture can be understood only as an historical growth. It is determined to a great extent by outer occurrences that do not originate in the inner life of the people.

It might be thought that these conditions did not prevail in early times, that primitive societies were isolated and that the laws governing their inner development may be learned directly from comparative studies of their cultures. This is not the case. Even the simplest groups with which we are familiar have developed by contact with their neighbors. The Bushman of South Africa has learned from the Negro; the Eskimo from the Indian; the Negrito from the Malay; the Veddah from the Singhalese. Cultural influences are not even confined to close neighbors; wheat and barley traveled in early times over a large part of the Old World; Indian corn over the two Americas.

If we find that the legal forms of Africa, Europe, and Asia are alike and different from those of primitive America, it does not follow that the American forms are more ancient, that the American and Old World forms represent a natural sequence, unless an actual, necessary order of the development can be demonstrated. It is much more probable that by cultural contact the legal forms of the Old World have spread over a wide area.

It is more than questionable whether it is justifiable to construct from a mere static examination of cultural forms the world over an historical sequence that would express laws of cultural development. Every culture is a complex growth and, on account of the intimate, early associations of people inhabiting large areas, it is not admissible to assume that the accidental causes that modify the course of development will cancel one another and that the great mass of evidence will give us a picture of a law of the growth of culture.

I am far from claiming that no general laws relating to the growth of culture exist. Whatever they may be, they are in every particular case overlaid by a mass of accidents that were probably much more potent in the actual happenings than the general laws.

We may recognize definite, causally determined relations between the economic conditions of a people and the size of population. The number of individuals of a hunting tribe inhabiting a particular territory is obviously limited by the available amount of game. There will be starvation as soon as the population exceeds the maximum that may be maintained in an unfavorable year. If the same people develop agriculture and the art of preserving a food supply for a long period, a denser population is possible and, at the same time, each individual will have more leisure and there will be a greater number of individuals enjoying leisure. Under these conditions the population is liable to increase. We may perhaps say that complexity of culture and absolute number of individuals constituting a population are correlated. Whether this development actually occurs in a given population is an entirely different question.

As another example of what might be called a social law I mention the re-interpretation of traditional behavior and belief. It may be claimed as a general rule that interpretations of customs and attitudes do not agree with their historical origins but are based on the general cultural tendencies of the time when the interpretation is given. Examples have been given before (see pp. 164 et seq.).

Still another example of what might be described as a social "law" will not be amiss. Important actions, when accompanied by difficulty of execution and likelihood of failure, or those involving danger, give rise to a variety of types of ceremonial behavior. The making of a canoe is often an act of great ceremonial importance, as in Polynesia, or is accompanied by superstitious beliefs and practices, as on the Northwest coast of America. Hunting and fishing on which the sustenance of the people depends, agricultural pursuits, herding, and war expeditions, are almost always connected with more or less elaborate ceremonials and complex beliefs, on the whole the more so the more deeply success or failure affect the life of the people. We may recognize an expression of the same "law" in the formal celebrations with which we like to accompany the achievement of great technical undertakings, the completion of the education of the young, or the opening of an important assembly.

Generalizations of this type are possible, but they do not enable us to predict the actual happenings in a specific culture. Neither do they allow us to lay down general laws governing the course of the historical development of culture.

When we try to apply the results of anthropological studies to the problems of modern life, we must not expect results parallel to those obtained by controlled experiments. The conditions are so complex that it is doubtful whether any significant "laws" can be discovered. There are certain tendencies in social behavior which are manifest; but the conditions in which they are active are controlled by accident, in so far as the varied activities of society and its relation to the outer world are logically unrelated. To give only one example: the technical development of electricity depended upon purely scientific work. The scientific discoveries depended upon the general advance of physics and upon purely theoretical interests. They were seized upon by the tendency of our times to exploit every discovery technically. The modifications of our lives brought about by the use of the telephone, radio, Roentgen rays, and the many other inventions are so little related to the scientific discovery itself that in relation to them it plays the rôle of an accident. If some of the discoveries had been made at another time their effect upon our social life might have been quite different. Thus every change in one aspect of social life acts as an accident in relation to others only remotely related to it.

For these reasons anthropology will never become an exact science in the sense that the knowledge of the status of a society at a given moment will permit us to predict what is going to happen. We may be able to *understand* social phenomena. I do not believe that we shall ever be able to *explain* them by reducing one and all of them to social laws.

These viewpoints must be borne in mind when we try to approach the problems of cultural progress. They may also help us in a critique of some of the theories on which modern social aspirations are based.

The rapid development of science and of the technical application of scientific knowledge are the impressive indications of the progress of modern civilization.

An increase in our knowledge and in the control of nature, an addition of new tools and processes to those known before may well be called progress, for nothing need be lost, but new powers are acquired and new insight is opened. Much of the increase in knowledge is, at the same time, elimination of error and in this sense also represents a progress. In the acquisition of new methods of controlling the forces of nature no qualitative standard is involved. It is a quantitative increase in the extent of previous achievements. In the recognition of earlier errors our standard is truth; but at the same time the recognition of error implies more rational, often useful conclusions. In all these acquisitions a process of reasoning is involved. The achievements are a result of intellectual work extending over ever-widening fields and increasing in thoroughness.

The discovery of methods of preserving food, the invention of manifold implements of the chase and of tools for manufacture; of clothing, shelter, and utensils for everyday life; the discovery of agriculture and the association with animals that led to their domestication; the substitution of metals for stone, bone, and wood; all these are rungs on the long ladder that led to our modern inventions, which are now being added to with overwhelming rapidity.

Knowledge has been increasing apace. The crude observation of nature taught man many simple facts —the forms and habits of animals and plants, the courses of the heavenly bodies, the changes of weather and the useful properties of materials, of fire and of water.

A long and difficult step was taken when the acquired knowledge was first systematized and conscious inquiry was attempted intended to expand the boundaries of knowledge. In early times imagination was drawn upon to supply the causal links between the phenomena of nature, or to give teleological explanations that satisfied the mind. Gradually the domain for the play of imagination has been restricted and the serious attempt is being made to subject imaginative hypotheses to the close scrutiny of observation.

Thus we may recognize progress in a definite direction in the development of invention and knowledge. If we should value a society entirely on the basis of its technical and scientific achievements it would be easy to establish a line of progress which, although not uniform, leads from simplicity to complexity.

Other aspects of cultural life are not with equal ease brought into a progressive sequence.

This may be illustrated by the changes in cultural life effected by progress in technical knowledge and skill. Primitive tribes who must devote all their energies and all their time to the acquisition of the barest necessities of life have not produced much that would help towards the enjoyment of life. Their comforts, social pleasures, art products, and ceremonials are cramped by their daily needs. These begin to flourish when the conditions of life allow leisure. A comparison of the hard life of the Fuegians, Eskimo, Australians, and Bushmen with that of people who command an abundance of food and prolonged periods of rest from procuring necessary supplies, shows the effect of leisure upon cultural life. The wealth of products of the African Negroes, the time at their disposal for ceremonial and social functions, are based on their comparative freedom of care for their everyday sustenance. Fishermen, like those of the North Pacific coast of America, who enjoy seasons of rest during which they live on stored provisions, have developed a complex art and a social and ceremonial life full of interest to themselves. Abundance of food has enabled the Melanesians to develop a rich inner life. Everywhere among primitive people leisure and enrichment of culture go hand in hand, for with leisure develop new needs, and new needs create new inventions. But leisure alone is not sufficient. Unless the individual participates in a multiplicity of cultural activities, if his life is restricted within a narrow compass, leisure is unprofitable. In primitive society the participation of everyone in tribal life creates the condition for a useful employment of leisure. Where a leisure class is created and part of the people are compelled to drudge for them, the leisure class alone may profit from their more favorable conditions.

It is a reproach to our civilization that we have not learned to utilize the vastly increased leisure in the way done by primitive man. Until recently the intensity of technical activity which creates an ever-increasing desire for physical comforts and conveniences used to make such demands upon the time of all individuals that for the majority leisure was much restricted. Nevertheless the time required for their manual labor has been much reduced during the last century. In recent times the ease of production by mechanical means and the rationalization of production, together with establishment of ever new centers of production without a corresponding increase of centers of consumption, have created a condition in which there is ample leisure, but leisure so distributed that part of the people are engaged in feverish activity while many others stand aside, outside of the streams of production and therefore unable to make use of the enforced leisure, not contributing new cultural values, but a dead weight on human progress.

Thus the advance in technical knowledge, not accompanied by corresponding social adjustment of the distribution of leisure, has led to a waste of human energy that might contribute to the enjoyment of life.

Primitive life shows that leisure enriches human life, at least as long as all actively participate in the production of cultural values. Among them leisure of all is often obtained by the seasonal rest, enabling each individual to participate in the social life of the tribe.

The impoverishment of the masses brought about by our unfortunate distribution of leisure is certainly no cultural advance, and the term "cultural progress" can be used in a restricted sense only. It refers to increase of knowledge and of control of nature.

It is not easy to define progress in any phase of social life other than in knowledge and control of nature.

It might seem that the low value given to life in primitive society and the cruelty of primitive man are indications of a low ethical standard. It is quite possible to show an advance in ethical *behavior* when we compare primitive society with our own. Westermarck and Hobhouse have examined these data in great detail and have given us an elaborate history of the evolution of moral ideas. Their descriptions are quite true, but I do not believe that they represent a growth of moral *ideas,* but rather reflect the same moral ideas as manifested in different types of society and taking on forms varying according to the extent of knowledge of the people.

We must bear in mind the meaning of a clean way of living, of practising all the virtues demanded by tradition. Sexual purity, avoidance of contamination by anything that may be impure, determine the specific code of almost every tribe. The transgression of any of the social customs that have a strong emotional value in the life of the people is considered a sin. The history of what is considered in our civilization a sin or a punishable wrong shows that the range of these concepts varies with our more or less rational attitude towards the particular aspect of our social life. Heterodoxy was a crime, atheism unpardonable; the breach of food taboos was not easily condoned; work on Sunday a sin. Abnormal sexual behavior was and still is punished, although we begin to recognize it as the result of biological factors, so that it is being considered more as a pathological state than as a punishable wrong. The restriction of sexual relations to a status authorized by church or State shows considerable weakening. The avoidance of all these sins: piety which included the observance of restrictions hallowed by the Church; sexual purity until the time when Church and State permit sexual intercourse, are analogous in character to the "sins" found in primitive society. The lives of the martyrs of all times who died for convictions that ran counter to the social laws of their times illustrate the intensity with which a breach was felt as an unpardonable sin. Native life abounds in analogous examples, both in actual occurrences and in novelistic tales. The transgressors of taboos—the unbeliever as well as the careless sinner—are punished by the supernatural powers and by society. To break a social incest law, to ignore a taboo, to omit a prescribed purification, are unpardonable sins. It is fairly clear, that in regard to all these cases it is rather a question of the advance of knowledge which makes the traditional regulation of life obselete, than a change in a feeling for ethical obligation that brings about changes in ethical behavior.

The question is rather whether there are certain fundamental ethical attitudes that are manifest in varying forms in all branches of mankind.

If we restrict our considerations to the closed society to which an individual belongs we do not find any appreciable difference in principles of morality. We have seen at another place that in a closed society without differentiation in rank, in theory at least, an absolute solidarity of interest and the same moral obligation of altruistic behavior are the ideal code, the same as among ourselves. The behavior towards the slave or to members of alien societies may be cruel. There may be no regard for their rights. The obligations within the society are binding. The prevailing idea of a fundamental, even specific difference between the members of the closed society and outsiders hinders the development of sympathetic feeling.

We consider it our right to kill criminals dangerous to society, to kill in self-defense and in war. We also kill animals for the mere pleasure of hunting and the excitement of the chase. Exactly the same rules prevail in primitive society. They give a different impression, because crime, self-defense, war, and the killing of animals have not the same meaning as among ourselves. A breach of the laws regulating marriage may be considered a heinous crime endangering the existence of the whole community because it calls forth the ire of supernatural powers; an apparently slight breach of good manners may be a deadly insult. Supposed witchcraft may be criminal or may entitle the person who believes himself endangered by it, as a matter of self-defense, to kill the offender. War may not be initiated by the formalities ordinarily enjoined by modern international usage—although often enough disregarded, if necessity or self-interest make it desirable—but may be based on a hostility between groups that blazes up on the slightest provocation and without warning, permitting what we should call basest treachery.

It is true that in the life of primitive man revenge as a right and a duty is keenly felt and that its form is much more cruel than our ethical standards would permit. In judging the psychological causes of this difference we must consider the infinitely greater hazards of life in primitive society. The weather, the dangers of the chase, attacks of wild animals or of enemies make life much more precarious than in civilized communities and dull the feeling for suffering. The thoughtless pleasure that children feel in tormenting animals and cripples, an expression of their inability to identify their own mental processes with those of others, is quite analogous to the actions of primitive man. The significance of this attitude will best be understood when we compare our feeling of sympathy for animal suffering with that of the Hindu. While we kill animals that we need for food, albeit without inflicting unnecessary suffering, all life is sacred to the Hindu. We claim the right to kill animals which we need; the Hindu extends the right to live over all his fellow creatures.

It might seem that the virtue of forgiving wrongs is entirely alien to primitive culture, for retaliation is almost always considered a duty. We recognize forgiving as a virtue, the more so since it is not always practised. Nevertheless we are still far from appreciating that legal punishment is rather a revenge of society than either protection against a dangerous criminal or an attempt at re-education. I believe the apparent absence of forgiving in primitive society is, like primitive cruelty, related in part to the precariousness of existence and the consequent necessity of self-protection, in part to the enmity between closed societies in which, under the pressure of public opinion, the individual is compelled to participate. Forgiving and tolerance between closed societies is difficult to find even in our civilization; witness the relations between nations, party and denominational quarrels, or those arising from keen competition in business affairs. It is necessary to examine primitive life attentively to see that the idea of forgiving wrongs as something praiseworthy is not absent. Every now and then it appears in folk tales clearly recognized as a desirable attitude. On the Northwest Coast of America the deserted boy who becomes rich saves from starvation his tribe who had no pity on him, although in other tales he retaliates on those who were the instigators of his misfortunes. Among the Pueblos children deserted by their own parents save them as soon as they are convinced of their repentance.

We must compare the code of primitive ethics with our own ethics and primitive conduct with our own conduct. It may safely be said that the code, so far as relations between members of a group are concerned, does not differ from ours. It is the duty of every person to respect life, well-being and property of his fellows, and to refrain from any action that may harm the group as a whole. All breaches of this code are threatened with social or supernatural punishment.

When the tribe is divided into small self-contained groups and moral obligations of the individual are confined to the group members, a state of apparent lawlessness may result. When the tribe forms a firm unit, the impression of peaceful quiet, more closely corresponding to our own conditions, is given. An example of the former kind is presented by the tribes of northern Vancouver Island, which are each divided into a considerable number of clans or family groups of conflicting interests. Solidarity does not extend beyond the limits of the clan. For this reason conflicts between clans are rather frequent. Harm done to a member of one clan leads to clan feuds.

The distinction between members of a group and outsiders persists in modern life, not only in everyday relations but also in legislation. Every law discriminating between citizens and foreigners, every protective tariff that is by its nature hostile to the foreigner is an expression of a double ethical standard, one for fellows, the other for outsiders.

The duty of self-perfection has developed in modern society, but is apparently absent in more primitive forms of human life. The irreconcilable conflicts of valuations that are characteristic of our times and to which we referred previously are in part absent because in simple societies a single standard of behavior prevails. We have referred to the freedom of the Eskimo of human control and have seen that, nevertheless, he is hemmed in on all sides by the narrowness of his material culture, his beliefs and traditional practices. There is no group known to him that possesses different standards, that presents the problem of choice between conflicting cultural alternatives that beset our lives, although conflicts based on different aspects of his own culture may arise. We have also referred to the social development of the child in Samoa where the lack of stratification into groups of decidedly distinct ideals makes it exceedingly difficult for new types of thought to develop. It does occur every now and then that a person does not fit temperamentally into his culture, as for instance a timid, unambitious nobleman or an aggressive, ambitious commoner among the Northwest Coast Indians; but these cases are as a rule rare and it is difficult for the individual to impress his qualities upon his environment. Thus it happens that the ethical duties that we feel towards ourselves, that in some strata of our society set the duty of self-perfection infinitely higher than that of service to the community, seem lost in the simple endeavor of every person to come up to the standards of his society.

The actual conduct of man does not correspond to the ethical code, and obedience depends upon the degree of social and religious control. Among ourselves actions opposed to the ethical code are checked by society, which holds every single person responsible for his actions. In many primitive societies there is no such power. The behavior of an individual may be censured, but there is no strict accountability, although the fear of supernatural punishment may serve as a substitute.

There is no evolution of moral ideas. All the vices that we know, lying, theft, murder, rape, are discountenanced in the life of equals in a closed society. There is progress in ethical conduct, based on the recognition of larger groups which participate in the rights enjoyed by members of the closed society, and on an increasing social control.

It is difficult to define progress in ethical ideas. It is still more difficult to discern universally valid progress in social organization, for what we choose to call progress depends upon the standards chosen. The extreme individualist might consider anarchy as his ideal. Others may believe in extreme voluntary regimentation; still others in a powerful control of the individual by society or in subjection to intelligent leadership. Developments in all these directions have occurred and may still be observed in the history of modern States. We can speak of progress in certain directions, hardly of absolute progress, except in so far as it is dependent upon knowledge which contributes to the safety of human life, health, and comfort.

Generally valid progress in social forms is intimately associated with advance in knowledge. It is based fundamentally on the recognition of a wider concept of humanity, and with it on the weakening of the conflicts between individual societies. The outsider is no longer a person without rights, whose life and property are the lawful prey of any one who can conquer him, but intertribal duties are recognized. However these are developed, whether the tribe wishes to avoid the retaliation of neighbors, or whether friendly relations are established by intermarriage or in other ways, the intense solidarity of the tribal unit and its subdivisions is liable to break down.

The important change of attitude brought about by this expansion is a weakening of the concept of a status into which each person is born.

The history of civilization demonstrates that the extent to which the status of a person is determined by birth or by some later voluntary or enforced act has been losing in force. In primitive societies of complex structure the status of a person as a member of a clan, of an age group, of a society, is often absolutely determined and involves unescapable obligations. Laws of intermarriage are determined by the status of a person in his or her family or hereditary tribal division and prevent the free choice of mates. Obligations and privileges may also vary according to the particular family or division in which a person is born. In East Africa agricultural classes and herders are hereditary tribal divisions. Chieftaincy in Polynesia and Africa and in many other parts of the world is hereditary in genealogical lines. All hereditary privileges belong to this class and these continue even in our times. Royal succession, the entailment of estates descending in family lines, compulsory laws prescribing the amounts to which lineal heirs are entitled are expressions of the recognition of a status into which a person is born. A status may also be innate, although not determined by family bonds. In South Africa a person who is believed to have the quality of witchcraft can never lose this status because it is inborn.

The status of age and sex plays an important rôle where rigidly organized age societies exist. Among some Negro tribes the boy is inducted into a group of boys of his own age who retain throughout life the status of a society to which each member is bound by social obligations. In Australia the aged form a group of recognized authority. In countries maintaining armies with obligatory service the male citizen has a definite status in regard to military duties, depending upon his age. This is true of the Zulu as well as of the French or Poles.

Membership in societies may also determine the status of a person. Sometimes the status is permanent, sometimes it may be changed with the consent of the community, expressed by some public, often religious act. In most primitive societies a priest cannot lay down the duties he has undertaken. The secret societies of West Africa which exert political powers give a permanent status to their members.

In earlier times, among ourselves, the status of the nobleman, of the serf, even of a member of a guild, was fixed by birth; that of the priest by authority of the Church. For most of us there are still two forms of status that entail serious obligation and that persist unless the status is changed by consent of the State. These are citizenship and marriage. The latter status shows even now strong evidence of weakening. In the sense of loss of fixity of status the freedom of the individual has been increasing.

The multitude of forms found in human society as well as observations on the variability of human types throw important light upon modern political questions, particularly upon the demand for equality, upon sexual relations and upon the denial of the right to individual property.

Anatomy, physiology, and psychology of social groups demonstrate with equal force that equality of all human beings does not exist. Bodily and mental ability and vigor are unevenly distributed among individuals. They also depend upon age and sex. Even in the absence of any form of organization which implies subordination, leadership develops. Eskimo society is fundamentally anarchical because nobody is compelled to submit to dictation. Nevertheless the movements of the tribe are determined by leaders to whose superior energy, skill, and experience others submit. The man, the provider of the family, determines the movements of the household and his wives and dependents follow.

It depends upon historical conditions to what extent the powers of a leader may be developed. In early times monarchical institutions spread over a large part of the Old World, democratic institutions over the New World. It is common to all forms of political organization that wherever communal work has to be undertaken, recognized leaders spring up.

Among the North American Indians who were averse to centralized political control, the buffalo hunt necessitated strict police regulations to which the tribe had to submit, because disorganized, individual hunting would have endangered the tribal food supply. The hunt and war in particular require leadership. How far each individual must submit to leadership depends upon the complexity of organization, upon the necessity of joint action, and upon conflicts arising from individual occupations.

The assumption that all leadership is an aberration from the primitive nature of man and an expression of individual lust for power cannot be maintained. We have pointed out repeatedly that man is a gregarious being, living in closed societies, and that new closed societies are always springing up. Almost all closed societies of animals have leaders and in many cases a definite order of rank may be observed. A typical case is the organization of a chicken yard in which a definite order of rank prevails. The first hen pecks the second, the second the third, and so on to the last which is pecked by all. The order is disturbed only if one chicken revolts and succeeds in overcoming its superior whose place it then takes. Other examples are the herds of mammals which have their scouts and watchmen and which protect themselves in orderly formation. It seems improbable that conditions were different in the primitive horde of man.

Observations on primitive society throw an interesting light upon the relation of the sexes. We find everywhere a clear distinction between the occupations of man and woman. Most women, being encumbered throughout a large part of their mature lives by the care of young children, are tied to the home more rigidly than the men. They are hindered in their mobility and for this reason more than for anything else they cannot participate in the strenuous life of the hunter and warrior. Here also a comparison with the life forms of gregarious animals is useful, for division of duties according to sex is not unusual. In some species the males are protectors of the herd, in other cases the females.

The domestic occupations of the home do not necessarily preclude women from active participation in the higher cultural life of the tribe. Owing to the skill attained in their varied technical activities they are in some cases creative artists, while the men who devote themselves to the chase do not participate to any extent in artistic production. Where a more complex economic system prevails in which wealth depends upon the management and care of the produce secured by the members of the household, her influence in social or even political matters may be important. She is not necessarily excluded from religious activities and acts as shaman or priestess.

Since among primitive tribes unmarried women are all but unknown, the position of womanhood is practically determined by the limitations imposed upon all by child-bearing and care of children.

Among primitive tribes the mortality of infants is high, and the intervals between births are correspondingly short. With the modern decrease in infant mortality, voluntary reduction of the number of children and the increasing number of unmarried women, the movements of many women have become freer and one of the fundamental causes of the differentiation between the social positions of men and of women has been removed. It is not by any means solely economic pressure that has led to the demand for wider opportunities and equality of rights of men and women, but the removal of the limitations due to child-bearing that has given to woman the freedom of action enjoyed by man.

The cultural values produced by woman in primitive society make us doubt the existence of any fundamental difference in creative power between the sexes. We rather suspect that the imponderable differences in the treatment of young children, the different attitudes of father and mother to son or daughter, the differentiation of the status of man and woman inherent in our cultural tradition, outweigh any actual differences that may exist.

In other words, the creative power and independence of man and of woman seem to me largely independent of the physiologically determined differences in interests and character. The danger in the modern desire of woman for freedom lies in the intentional suppression of the functions connected with child-bearing that might hinder free activity. Society will always need a sufficient number of women who will bear children and of those who are willing to devote themselves lovingly to their upbringing.

Marriage is another aspect of the relation between the sexes upon which light is thrown by the study of foreign cultures. The customs of mankind show that permanent marriage is not based primarily on the permanence of sexual love between two individuals, but that it is essentially regulated by economic considerations. Formal marriage is connected with transfer of property. In extreme cases the woman herself is an economic value that is acquired, although she may not become the property of her husband in the sense that he can dispose of her at will without interference of her own family or herself.

Occasional sexual relations between man and woman are of a different order and are among many tribes permitted or even expected. In other cases girls are carefully guarded and illicit sexual intercourse is severely punished.

A religious sanction of marriage exists in hardly any primitive tribe. Strict monogamy does occur in rare cases and suggests that the sexual relations in earliest times were not of uniform character in all parts of the world. The binding elements in marriage are considerations of property in which the children who add potential strength to the family are included.

It seems likely that our view of marriage developed from this earlier stage by reinterpretation.

In a well balanced family with competent parents, permanence of matrimonial union is undoubtedly best adapted to the wholesome development of the individual and of society. But not all families are well balanced and competent, and permanence of affection is not universal. On the contrary, almost all societies illustrate fickleness of affection and instability of unions among young people. Unions become fairly stable only in old age, when the sexual passions have abated. Instability is found as much in modern civilization as in simpler societies. Man is evidently not an absolutely monogamous being.

The efforts to force man into absolute monogamy have never been successful and the tendency of our times is to recognize this. The increasing ease of divorce which has been carried furthest in Mexico and Russia is proof of this. Equally significant are the endeavors to ease the unenviable position of the unmarried mother, the attempts to lift the undeserved stigma from the illegitimate child, and the claims for a single standard of sexual ethics for man and woman.

The anthropologist may not be able to propose on the basis of his science the steps that should be taken to remedy the hypocrisy that attaches to the general treatment of sexual relations without unduly encouraging the light-hearted breaking of the marriage bond. He can only point out that the traditional point of view of absolute continence until a monogamic marriage is contracted is not enforceable, because it runs counter to the nature of a large part of mankind. In many cases it is accepted and followed like other social standards, but not without giving rise to severe crises.

It is interesting to investigate the concept of property in simple tribes. We do not know of a single tribe that does not recognize individual property. The tools and utensils which a person makes and uses are practically always his individual property which he may use, loan, give away or destroy, provided he does not damage the life of his household by doing so. An Eskimo man who would destroy his kayak and hunting outfit would make himself and his family dependent upon the industry of others; the Eskimo woman who would destroy her cooking utensils or her clothing would deprive the family of valuable property which could not be replaced without the help of her husband or other men. In this sense the control of their property is not absolutely free. Any economic theory that does not acknowledge these facts runs counter to anthropological data.

The concept of property in natural resources is of a different character. Except in the rare cases of truly nomadic peoples, the tribe is attached to a definite geographical area which is its property in so far as foreigners who would try to utilize it are considered as intruders. In simpler societies tribal territory and all its resources belong to the community as a whole; or when the tribe consists of subdivisions the tribal territory may be subdivided among them, and mutual encroachments will not be permitted. In most cases it is not necessary to develop the natural resources by labor and the supply is ample for the needs of the people. Stone, shell, wood, or pure metal for manufactures are more or less easily obtained. When preparatory labor is involved in making the products of nature available property rights develop. The African who clears the woods, and plants and cultivates his garden, has property rights to the soil until he deserts it and allows it to revert to its original wild state. The Northwest Coast Indian who builds a fish weir at a favorable place considers it his property. The greater the amount of labor bestowed by an individual, family, or clan upon the exploitation of a given piece of land or water, the more are we likely to find the concept of personal, family, or clan property in the ground and its products. In places where permanent houses are built a similar relation may develop to the building site. Herding which requires constant attention to the welfare of the animals on the part of the herder establishes a close connection between the two and, unless feudal conditions prevail in which the herd is the property of an overlord, the herd is the property of the herder. In all these cases conflicts are liable to develop. In Africa and Melanesia the rights to the use of land are regulated; sometimes the use of abandoned land may pass from one hand to another, while the use of trees may be retained by the former owner. Among the Indians quarrels are common in regard to the right to use fish weirs, or even places at which to erect fish weirs. Among herders, cattle stealing is a common source of local feuds. The conflict between the feeling that personal control of natural resources infringes upon the interests of the community as a whole arises at an early time. In modern times when the development of natural resources by a powerful person or group of persons has become necessary because profitable exploitation requires scientific knowledge, the use of machinery, of ample means and of methods of controlling the wide distribution of the products; and when those who are in control claim the resources as their personal property because they are the means of putting them into use the conflict between the property claims of the community and of individuals has reached its highest grade.

It is not possible to follow in the brief compass of these remarks the variety of concepts of property that develop from primitive control: the centralization of ownership in the hands of a favored class or of individuals, and the privileges that grow up with increasing complexity of society.

Theories of the growth of culture have been built, based on the assumption of the determining influence of single causes. Most important among these are the theories of geographical and economic determinism.

Geographical determinism means that geographical

environment controls the development of culture; economic determinism that the economic conditions of life shape all the manifestations of early culture and of complex civilization.

It is easy to show that both theories ascribe an exaggerated importance to factors that do play an important part in the life of man, but that are each only one of many determinant elements.

The study of the cultural history of any particular area shows clearly that geographical conditions by themselves have no creative force and are certainly no absolute determinants of culture.

Before the introduction of the horse the western American prairies were hardly inhabited, because the food supply was uncertain. When the Indians were supplied with horses their whole mode of life changed, because buffalo hunting became much more productive and the people were able to follow the migrating herds of buffalo. Many tribes migrated westward and gave up agriculture. When the White man settled on the prairies, life was again different. Agriculture and herding were adapted to the new environment. According to the type of culture of the people who occupied the prairies, these played a different rôle. They compelled man to adapt his life to the new conditions and modified the culture. The environment did not create a new culture.

Another example will not be amiss. The Arctic tundra in America and Asia has about the same character. Still the lives of the Arctic Indians and Eskimos and that of the tribes of Siberia are not the same. The Americans are exclusively hunters and fishermen. The Asiatics have domesticated reindeer. The environment has not the same meaning for the hunter and for the herder; but herding was not invented owing to the stress of environment. It is a type of Asiatic culture that takes a particular form in the Arctic climate.

When the principal trade routes from Europe to the East crossed the Mediterranean Sea and vessels were of moderate size, the distribution of trade centers, of sea routes and of available harbors was quite different from that found in later times, when, owing to shifts in political and cultural conditions, to new discoveries, new demands, and in modern times, to larger vessels, the same environment brought about new alignments, decay to once flourishing cities, and increased importance to others.

The error of the theory of geographic determinism lies in the assumption that there are tribes on our globe without any culture, that must learn to adapt themselves to the environment in which they live. We do not know of any tribe without some form of culture and even in the times of the older stone age, perhaps 50,000 years ago, this condition did not exist. The environment can only act upon a culture and the result of environmental influences is dependent upon the culture upon which it acts. Fertility of the soil has nowhere created agriculture, but when agriculture exists it is adapted to geographical conditions.

Presence of iron ore and coal does not create industries, but when the knowledge of the use of these materials is known, geographical conditions exert a powerful influence upon local development.

Geographical conditions exert a limiting or modifying power, in so far as available materials, topographical forms, and climate compel certain adjustments, but many different types of culture are found adjusted to similar types of environment.

The error that is often committed is similar to the one that has for a long time made experimental psychology unproductive. There is no society without some type of culture, and there is no blank mind upon which culture,—or bringing up of the individual,—has left no impress. An immediate reaction of the mind to a stimulus depends not alone upon the organization of the mind and the stimulus, but also upon the modifications that the mind has undergone, owing to its development in the setting of a culture.

Economic determinism is open to the same objections. The theory is more attractive than geographic determinism because economic conditions are an integral part of culture and are closely interwoven with all its other aspects. In our life their influence makes itself felt in the most varied forms and modern civilization cannot be understood without constant attention to its economic background.

Nevertheless it would be an error to claim that all manifestations of cultural life are determined by economic conditions. The simplest cultural forms prove this. There are many tribes of hunters and fishermen whose economic life is built up on similar foundations. Nevertheless they differ fundamentally in customs and beliefs. African Bushmen and Australian Aborigines; Arctic Indians and some of the river tribes of Siberia; Indians of Alaska, Chile and the natives of the island of Saghalin in eastern Asia are comparable, so far as their economic resources are concerned. Still their social organization, their beliefs and customs are diverse. There is nothing to indicate that these are due to economic differences; rather the use of their economic resources depends upon all the other aspects of cultural life.

Even the differences in the status of man and woman are not primarily economic. They are rather due to the differences in the physiological life of man and of woman. Based on this there is a difference in occupation, in interests and in mental attitude. These in turn produce economic differentiation, but the economic status is not the primary cause of the status of man and woman.

We may observe here that what is an effect of differentiation, becomes a cause of further differentiation. This relation may be observed in all specific phenomena of nature. A valley has been formed as the effect of erosion. It is the cause that in the further action of erosion the waters follow its course. Luxurious vegetation is the effect of a moist soil. It is the cause of retaining more moisture in the soil. A household performs joint work, and the joint work strengthens the unity of the household. Leisure obtained by the preservation of a plentiful supply of food stimulates invention, and the inventions give more leisure.

The interaction between the various forces is so intimate that to select one as the sole creative force conveys an erroneous impression of the process. It seems impossible to reduce the fundamental beliefs of mankind to an economic source. They arise from a variety of sources, one of which is the unconscious conceptualization of nature. The organization of the household is controlled in part by the size of the economic unit allowed by the food supply, in part by ties of association that are established by beliefs or habits so slightly related to economic conditions that it would require great ingenuity and a forced reasoning to reduce them to economic causes.

It is justifiable to investigate the intricate relations of economic life and of all the other numerous manifestations of culture, but it is not possible to rule out all the remaining aspects as dependent upon economic conditions. It is just as necessary to study economic life as dependent upon inventions, social structure, art, and religion as it is to study the reverse relations.

Economic conditions are the cause of many of these and they are with equal truth their effect. Social bonds and conflicts, concepts, emotional life, artistic activities are in their psychological and social origin only incompletely reducible to economic factors.

As geographical environment acts only upon a culture modifying it, so economic conditions act upon an existing culture and are in turn modified by it.

A final question must be answered. Can anthropology help to control the future development of human culture and well-being or must we be satisfied to record the progress of events and let them take their course? I believe we have seen that a knowledge of anthropology may guide us in many of our policies. This does not mean that we can predict the ultimate results of our actions. It has been claimed that human culture is something superorganic, that it follows laws that are not willed by any individual participating in the culture, but that are inherent in the culture itself. Some of the gradual changes referred to before might seem to support this view. The increase of knowledge, the freeing of the individual from traditional fetters, the extension of political units have proceeded regularly.

It seems hardly necessary to consider culture a mystic entity that exists outside the society of its individual carriers, and that moves by its own force. The life of a society is carried on by individuals who act singly and jointly under the stress of the tradition in which they have grown up and surrounded by the products of their own activities and those of their forbears. These determine the direction of their activities positively or negatively. They may proceed to act and think according to the transmitted patterns or they may be led to move in opposite directions. Occupation with a thought or an invention may lead on in different directions. Seen retrospectively they may appear like a predetermined growth.

The state of the society at a given moment depends upon the interactions of the individuals under the stress of traditional behavior. It is not the sum of the activities of the individuals; rather individuals and society are functionally related.

The forces that bring about the changes are active in the individuals composing the social groups, not in the abstract culture.

Here, as well as in other social phenomena, accident cannot be eliminated, accident that may depend upon the presence or absence of eminent individuals, upon the favors bestowed by nature, upon chance discoveries or contacts, and therefore prediction is precarious, if not impossible. Laws of development, except in most generalized forms, cannot be established and a detailed course of growth cannot be predicted.

All we can do is to watch and judge day by day what we are doing, to understand what is happening in the light of what we have learned and to shape our steps accordingly.

REFERENCES

In the following pages some of the more important literature is quoted on which are based the statements made in the text of the book:

P. 18. Claims of fundamental racial differences will be found in A. de Gobineau, *Essai sur l'inégalité des races;* Madison Grant, *The Passing of the Great Race;* Hans F. K. Günther, *Rassenkunde des deutschen Volkes;* Houston Stewart Chamberlain, *Die Grundlagen des XIX Jahrhunderts.* The opposite view is held by Th. Waitz, *Anthropologie der Naturvölker,* 2nd edition, Vol. 1, p. 381; Franz Boas, *The Mind of Primitive Man;* Friedrich Hertz, *Race and Civilization;* Ignaz Zollschan, *Das Rassenproblem.* An attempt at a critical review of the literature is contained in Frank H. Hankins, *The Racial Basis of Civilization,* and more fully, but with a complete misunderstanding of the meaning of racial and national traits in Théophile Simar, *Etude critique sur la formation de la doctrine des races.*

P. 26. The effect of inbreeding upon the variability of family lines has been discussed by F. Boas in the American Anthropologist N. S. Vol. 18 (1916), pp. 1 et seq., and by Isabel Gordon Carter, American Journal of Physical Anthropology, Vol. 11, pp. 457 et seq. See also Eugen Fischer, *Die Rehobother Bastards;* Ernst Rodenwaldt, *Die Mestizen auf Kisar,* Jena, 1928.

P. 28. The foreign elements in the Swedish nation have been discussed by Gustav Retzius and Carl M. Fürst in *Anthropologia Suecica*, p. 18.

P. 32. The reversion to the type of a population has been first discussed by Francis Galton, *Natural Inheritance;* and later on elaborated by Karl Pearson, *Mathematical Contributions to the Theory of Evolution,* III; Royal Society of London, 1896, A, pp. 253 et seq. The varying regression from the same parental type to distinct populations has been illustrated by F. Boas, in the Proceedings of the National Academy of Sciences, Vol. 14, pp. 496 et seq.

P. 35. The changes of stature in Europe are summarized in Rudolf Martin, *Lehrbuch der Anthropologie,* pp. 224 et seq., 1st edition. Edw. Ph. Mackeprang, *De vaernepligtiges legemshojde i Danmark,* Meddelelser om Danmarks Antropologi, Vol. 1, pp. 1 et seq. Comparisons of parents and their own children are given in F. Boas, *Changes in Bodily Form of Descendants of Immigrants,* pp. 28, 30.

P. 36. A discussion of the measures of the hand according to occupation is found in E. Brezina and V. Lebzelter, *Ueber die Dimensionen der Hand bei verschiedenen Berufen,* Archiv für Hygiene, Vol. 92, 1923; and Zeitschrift für Konstitutionslehre, Vol. 10, pp. 381 et seq.

P. 36. A brief discussion of the effect of the use of the limbs upon the form of the leg bones is found in Zeitschrift für Ethnologie, Anthropologie und Urgeschichte, Vol. 17 (1885), p. 253; and by L. Manouvrier in Bulletin de la Société d'Anthropologie, Paris, Series 3, Vol. 10, p. 128.

P. 39. Adolph H. Schultz, *Man as a Primate,* Scientific Monthly, Nov. 1931, pp. 385-412.

P. 42. Traits due to domestication have been studied particularly by Eugen Fischer. His results have been briefly summarized in his book, *Rasse und Rassenentstehung beim Menschen.* See also B. Klatt, *Mendelismus, Domestikation und Kraniologie,* Archiv f. Anthropologie N. S., Vol. 18, pp. 225-250.

P. 45. A summary of the literature relating to the study of the senses of races is found in Gustav Kafka, *Handbuch der vergleichenden Psychologie,* Vol. 1, pp. 163 et seq.

P. 45. Differences in basal metabolism were found by Francis G. Benedict, *The racial factor in metabolism.* Proceedings National Academy of Sciences, Vol. 11, pp. 342 et seq.

P. 46. The margin of safety has been discussed by S. J. Meltzer, *Factors of safety in animal structure and animal economy,* Journal of the American Medical Association, Vol. 48, pp. 655 et seq.; Science, New Series, Vol. 25, pp. 481 et seq.

P. 53. O. Klineberg, *An experimental study of speed and other factors in "racial" differences.* Archives of Psychology, No. 93. Dr. Klineberg has also shown by a study of different human types in Germany, France and Italy that the results of the intelligence tests do not agree with the distribution of races but depend on social conditions. *A study of psychological differences between "racial" and national groups in Europe.* Archives of Psychology, No. 132, 1931.

P. 53. Paul Roloff has found considerable differences in the ability to define terms, among different social classes of the same district. *Beihefte zur Zeitschrift für angewandte Psychologie,* Vol. 27, pp. 162 et seq.

P. 58. Melville J. Herskovits discusses the mental behavior of a socially uniform group of negroes and mulattoes in *The American Negro.*

Pp. 59, 60. Studies of human culture without any regard to race are, for instance, Edward B. Tylor, *Primitive*

Culture; Herbert Spencer, *Principles of Sociology;* F. Ratzel, *The History of Mankind.* Adolf Bastian's viewpoint has been analyzed by Th. Achelis, *Moderne Völkerkunde,* pp. 189 et seq.

P. 60. Gustav Klemm, *Allgemeine Cultur-Geschichte,* Leipzig, 1843-1852; Carl Gustav Carus, *Ueber ungleiche Befähigung der verschiedenen Menscheitstämme für höhen geistige Entwickelung,* Leipzig, 1849.

P. 63. Romain Rolland speaks of "race antipathy" in his *Jean Christophe,* 44th edition, Vol. 10, p. 23.

P. 65. The relations between Negroes, Indians, and Whites in Brazil were described to me by Mr. Rüdiger Bilden, those in Santo Domingo by Dr. Manuel Andrade.

P. 67. Local forms in animal life have been described by Friedrich Alverdes, *Tiersoziologie,* 1925. The social life of insects has been treated by William M. Wheeler, *Social Insects, Their Origin and Evolution,* 1928.

P. 74. The actual distribution of mixtures of Whites and Negroes has been described by Melville J. Herskovits in the book previously referred to. The United States Census is not a reliable source for the relative number of Mulattoes and full-blood Negroes.

P. 77. Marriage preferences of Negroes and Mulattoes have been discussed by Herskovits in the book just mentioned.

P. 81. See for this chapter, Carlton J. H. Hayes, *Nationalism,* and W. Mitscherlich, *Nationalismus.*

P. 84. Recently Sir Arthur Keith, in a book entitled, "Ethnos, or the Problem of Race," has taken the point of view that behavior is more important than physical type and he discusses at some length the ways in which capitalism and economics have "upset" nature's plan of evolution. There is no doubt that our modern cultural conditions have selective effects different from those of earlier times, and it is justifiable to ask in how far the physical character of a race may be determined by these factors. When, however, Sir Arthur Keith proceeds and practically identifies the mental behavior of a group of people with their racial characteristics, he completely deserts the basic biological aspect of race and substitutes for it a mingled group differing in mental characteristics and possibly also in physical characteristics. It is no longer a race but a group held together by social or political bonds. Anything that is said about such a group has no longer any bearing upon what can be said about a group of homogeneous biological descent.

P. 87. Owing to the economic advantages accruing to Indians who speak Spanish, there is a marked tendency in many Mexican villages to discourage the use of the native language. I have heard Indian mothers reprimand their children for speaking Indian.

P. 88. The descent of the Swedish nobility has been discussed by P. E. Fahlbeck, *Der Adel Schwedens*, Jena, 1903.

P. 109. As examples of attempts to correlate bodily form and pathological or psychological conditions I mention George Draper, *Human Constitutions*, and E. Kretschmer, *Köperbau und Charakter*.

P. 110. For examples of heredity of defective traits see R. L. Dugdale, *The Jukes*, and A. H. Estabrook, *The Jukes in 1915;* also H. H. Goddard, *The Kallikak Family;* A. H. Estabrook and C. B. Davenport, *The Nam Family;* F. H. Danielson and C. B. Davenport, *The Hill Folk.*

P. 122. C. Lombroso's views are set forth in many of his books, for instance, *L'uomo delinquente in rapporto all' antropologia; L'homme criminel.*

P. 123. One of the most thorough studies of criminals is that by C. Goring, *The English Convict.*

P. 128. A discussion of observations in Habit Clinics is

given in Blanche C. Weill, *The Behavior of Young Children of the Same Family*, Cambridge, 1928.

P. 133. E. Conradi, American Journal of Psychology, 16, 1905.

P. 133. K. S. Lashley, Journal of Animal Behavior, 3, 1913.

P. 133. O. Menghin, Weltgeschichte der Steinzeit.

P. 146. The influence of habit upon form has been discussed by Franz Boas, *Primitive Art*, Oslo, 1927, pp. 143 et seq.

P. 149. In F. Boas, *Handbook of American Indian Languages*, the forms of primitive languages are discussed. See also Edward Sapir, *Language*.

P. 155. See Ruth Benedict, *Psychological types in the cultures of the southwest*, Proceedings of the 23rd International Congress of Americanists, New York, G. E. Stechert, 1930, pp. 572-581; *Configuration of Culture in North America*, American Anthropologist, New Series, Vol. 34, pp. 1-27.

P. 159. A remarkable instance illustrating the conservatism of the Eskimo and their Arctic neighbors is contained in a phrase in their folklore. In East Greenland it is told of a man "who was so strong even in death, that he did not lie stretched out, but rested on his muscles (*i.e.*, of the buttocks and shoulders)"; Knud Rasmussen, *Myter og Sagn from Grönland*, Vol. 1, p. 272 (figure). The Chukchee of eastern Siberia tell of a strong man who had been killed: "He lay there, touching the ground merely with his calves, with his shoulder blades, and with the other fleshy parts of his body." W. Bogoras, *Chukchee Texts*, Publications of the Jesup North Pacific Expedition, Vol. 8, p. 98.

P. 159. Changes of ritual are occurring among the Kwakiutl Indians of Vancouver Island, who during the last seventy or eighty years, have constantly been adding new features and dropping old ones. Changes owing to the premature death of the keeper of a secret ritual must have occurred repeatedly among the Pueblo tribes. The best authenticated case is that of the Pueblo of Cochití where, owing to the death of the chief of ceremonies and rituals during the absence of his successor, the latter had only a fragmentary knowledge of his duties.

P. 162. The best described example of the origin of a mixed religion is that of the Ghost Dance by James Mooney in the 14th Annual Report of the Bureau of American Ethnology.

P. 162. T. T. Waterman, *The Snake Religion of Puget Sound,* Smithsonian Report, for 1922, pp. 499-507

P. 162. The restriction of inventiveness of the artist has been described in F. Boas, *Primitive Art.*

P. 162. The individuality in design has been described by Ruth Bunzel in *The Pueblo Potter,* New York, Columbia University Press, 1927, pp. 64-68.

P. 164. The significance of the Sun Dance among various tribes has been studied by Leslie Spier, in Anthropological Papers of the American Museum of Natural History, Vol. 16, pp. 451 et seq.

P. 165. The stability of ornamental form and the variety of interpretation are illustrated in F. Boas, *Primitive Art,* pp. 88 et seq. Examples of the varied interpretation of tales have been collected by T. T. Waterman, Journal of American Folk-Lore, Vol. 27, pp. 1 et seq.

P. 166. The theory of survivals has been set forth by Edward B. Tylor, *Primitive Culture,* Vol. 1, pp. 70 et seq.

P. 166. Some of the rules of forbidden and prescribed cousin marriages will be found in Robert H. Lowie, *Primitive Society.*

P. 168. Maria Montessori, in her *Pedagogical Anthropol-*

ogy, has tried to apply anthropometric data to educational problems.

P. 171. The periods of certain physiological stages have been described by F. Boas in *Remarks on the Anthropological Study of Children*, Transactions of the 15th International Congress of Hygiene and Demography.

P. 173. Observations on variations in development according to social classes are found in many places. See, for instance, references in H. Ploss, *Das Weib*, 2nd edition, Vol. 1, p. 228; also H. P. Bowditch, *The Growth of Children*, 8th Annual Report State Bureau of Health of Massachusetts; C. Roberts, *A Manual of Anthropometry*. See also the selected bibliography in D. A. Prescott, *The determination of anatomic age in school children and its relation to mental development.*

P. 173. Milo Hellman, *Nutrition, Growth and Dentition*, Dental Cosmos, January, 1923. The earlier maturity of girls appears least clearly in the development of the permanent dentition.

P. 176. The differences in bodily form of boys and girls have been pointed out by me at the place just mentioned and by Ruth O. Sawtell, *Sex Differences in Bone Growth of Young Children*, American Journal of Physical Anthropology, Vol. 12, 1928, pp. 293-302.

P. 176. A full treatment of the psychology of childhood in its relation to educational problems is contained in E. L. Thorndike, *Educational Psychology*.

P. 177. A comparison of Jewish and Northwest European children will be found in F. Boas, *The Growth of Children as Influenced by Environmental and Hereditary Conditions, School and Society*, Vol. 17, p. 305 et seq.

P. 179. The uncertainty of correlations between height and weight and undernourishment have been set forth by Louis I. Dublin and John C. Gebhart, *Do Height and Weight Tables Identify Undernourished Children?*, New

York Association for Improving the Condition of the Poor.

P. 183. The correlations between stature attained at a certain age and subsequent growth have been discussed by Clark Wissler, American Anthropologist, New Series, Vol. 5, pp. 81 et seq. A fuller discussion of this subject is in preparation.

P. 189. See descriptions of the killing of aged persons in Waldemar Bogoras, *The Chukchee,* Publications of the Jesup North Pacific Expedition, Vol. 8.

P. 190. The life of the adolescent girl in Samoa has been discussed by Margaret Mead, *Coming of Age in Samoa.*

P. 200. Robert S. Lynd and Helen Merrell Lynd, *Middletown,* New York, 1929.

P. 215. For a discussion of these problems and literature see Handwörterbuch der Soziologie, Stuttgart, 1931. Article by F. Dobretsberger, *Historische und sociale Gesetze,* pp. 211-221.

P. 220. E. Westermarck, *The Origin and Development of Moral Ideas;* L. T. Hobhouse, *Morals in Evolution.*

P. 225. Franz Boas, Keresan Texts, American Ethnological Society, Vol. VIII, Part I, p. 147.

A CATALOG OF SELECTED
DOVER BOOKS
IN ALL FIELDS OF INTEREST

A CATALOG OF SELECTED DOVER BOOKS IN ALL FIELDS OF INTEREST

THE CLARINET AND CLARINET PLAYING, David Pino. Lively, comprehensive work features suggestions about technique, musicianship, and musical interpretation, as well as guidelines for teaching, making your own reeds, and preparing for public performance. Includes an intriguing look at clarinet history. "A godsend," *The Clarinet,* Journal of the International Clarinet Society. Appendixes. 7 illus. 320pp. 5⅜ x 8½. 0-486-40270-3

HOLLYWOOD GLAMOR PORTRAITS, John Kobal (ed.). 145 photos from 1926-49. Harlow, Gable, Bogart, Bacall; 94 stars in all. Full background on photographers, technical aspects. 160pp. 8⅜ x 11¼. 0-486-23352-9

THE RAVEN AND OTHER FAVORITE POEMS, Edgar Allan Poe. Over 40 of the author's most memorable poems: "The Bells," "Ulalume," "Israfel," "To Helen," "The Conqueror Worm," "Eldorado," "Annabel Lee," many more. Alphabetic lists of titles and first lines. 64pp. 5³⁄₁₆ x 8¼. 0-486-26685-0

PERSONAL MEMOIRS OF U. S. GRANT, Ulysses Simpson Grant. Intelligent, deeply moving firsthand account of Civil War campaigns, considered by many the finest military memoirs ever written. Includes letters, historic photographs, maps and more. 528pp. 6⅛ x 9¼. 0-486-28587-1

ANCIENT EGYPTIAN MATERIALS AND INDUSTRIES, A. Lucas and J. Harris. Fascinating, comprehensive, thoroughly documented text describes this ancient civilization's vast resources and the processes that incorporated them in daily life, including the use of animal products, building materials, cosmetics, perfumes and incense, fibers, glazed ware, glass and its manufacture, materials used in the mummification process, and much more. 544pp. 6⅛ x 9¼. (Available in U.S. only.) 0-486-40446-3

RUSSIAN STORIES/RUSSKIE RASSKAZY: A Dual-Language Book, edited by Gleb Struve. Twelve tales by such masters as Chekhov, Tolstoy, Dostoevsky, Pushkin, others. Excellent word-for-word English translations on facing pages, plus teaching and study aids, Russian/English vocabulary, biographical/critical introductions, more. 416pp. 5⅜ x 8½. 0-486-26244-8

PHILADELPHIA THEN AND NOW: 60 Sites Photographed in the Past and Present, Kenneth Finkel and Susan Oyama. Rare photographs of City Hall, Logan Square, Independence Hall, Betsy Ross House, other landmarks juxtaposed with contemporary views. Captures changing face of historic city. Introduction. Captions. 128pp. 8¼ x 11. 0-486-25790-8

NORTH AMERICAN INDIAN LIFE: Customs and Traditions of 23 Tribes, Elsie Clews Parsons (ed.). 27 fictionalized essays by noted anthropologists examine religion, customs, government, additional facets of life among the Winnebago, Crow, Zuni, Eskimo, other tribes. 480pp. 6⅛ x 9¼. 0-486-27377-6

TECHNICAL MANUAL AND DICTIONARY OF CLASSICAL BALLET, Gail Grant. Defines, explains, comments on steps, movements, poses and concepts. 15-page pictorial section. Basic book for student, viewer. 127pp. 5⅜ x 8½. 0-486-21843-0

THE MALE AND FEMALE FIGURE IN MOTION: 60 Classic Photographic Sequences, Eadweard Muybridge. 60 true-action photographs of men and women walking, running, climbing, bending, turning, etc., reproduced from rare 19th-century masterpiece. vi + 121pp. 9 x 12. 0-486-24745-7

ANIMALS: 1,419 Copyright-Free Illustrations of Mammals, Birds, Fish, Insects, etc., Jim Harter (ed.). Clear wood engravings present, in extremely lifelike poses, over 1,000 species of animals. One of the most extensive pictorial sourcebooks of its kind. Captions. Index. 284pp. 9 x 12. 0-486-23766-4

1001 QUESTIONS ANSWERED ABOUT THE SEASHORE, N. J. Berrill and Jacquelyn Berrill. Queries answered about dolphins, sea snails, sponges, starfish, fishes, shore birds, many others. Covers appearance, breeding, growth, feeding, much more. 305pp. 5¼ x 8¼. 0-486-23366-9

ATTRACTING BIRDS TO YOUR YARD, William J. Weber. Easy-to-follow guide offers advice on how to attract the greatest diversity of birds: birdhouses, feeders, water and waterers, much more. 96pp. 5³⁄₁₆ x 8¼. 0-486-28927-3

MEDICINAL AND OTHER USES OF NORTH AMERICAN PLANTS: A Historical Survey with Special Reference to the Eastern Indian Tribes, Charlotte Erichsen-Brown. Chronological historical citations document 500 years of usage of plants, trees, shrubs native to eastern Canada, northeastern U.S. Also complete identifying information. 343 illustrations. 544pp. 6½ x 9¼. 0-486-25951-X

STORYBOOK MAZES, Dave Phillips. 23 stories and mazes on two-page spreads: Wizard of Oz, Treasure Island, Robin Hood, etc. Solutions. 64pp. 8¼ x 11. 0-486-23628-5

AMERICAN NEGRO SONGS: 230 Folk Songs and Spirituals, Religious and Secular, John W. Work. This authoritative study traces the African influences of songs sung and played by black Americans at work, in church, and as entertainment. The author discusses the lyric significance of such songs as "Swing Low, Sweet Chariot," "John Henry," and others and offers the words and music for 230 songs. Bibliography. Index of Song Titles. 272pp. 6½ x 9¼. 0-486-40271-1

MOVIE-STAR PORTRAITS OF THE FORTIES, John Kobal (ed.). 163 glamor, studio photos of 106 stars of the 1940s: Rita Hayworth, Ava Gardner, Marlon Brando, Clark Gable, many more. 176pp. 8⅜ x 11¼. 0-486-23546-7

YEKL and THE IMPORTED BRIDEGROOM AND OTHER STORIES OF YIDDISH NEW YORK, Abraham Cahan. Film Hester Street based on *Yekl* (1896). Novel, other stories among first about Jewish immigrants on N.Y.'s East Side. 240pp. 5⅜ x 8½. 0-486-22427-9

SELECTED POEMS, Walt Whitman. Generous sampling from *Leaves of Grass*. Twenty-four poems include "I Hear America Singing," "Song of the Open Road," "I Sing the Body Electric," "When Lilacs Last in the Dooryard Bloom'd," "O Captain! My Captain!"–all reprinted from an authoritative edition. Lists of titles and first lines. 128pp. 5³⁄₁₆ x 8¼. 0-486-26878-0

SONGS OF EXPERIENCE: Facsimile Reproduction with 26 Plates in Full Color, William Blake. 26 full-color plates from a rare 1826 edition. Includes "The Tyger," "London," "Holy Thursday," and other poems. Printed text of poems. 48pp. 5¼ x 7. 0-486-24636-1

THE BEST TALES OF HOFFMANN, E. T. A. Hoffmann. 10 of Hoffmann's most important stories: "Nutcracker and the King of Mice," "The Golden Flowerpot," etc. 458pp. 5⅜ x 8½. 0-486-21793-0

THE BOOK OF TEA, Kakuzo Okakura. Minor classic of the Orient: entertaining, charming explanation, interpretation of traditional Japanese culture in terms of tea ceremony. 94pp. 5⅜ x 8½. 0-486-20070-1

FRENCH STORIES/CONTES FRANÇAIS: A Dual-Language Book, Wallace Fowlie. Ten stories by French masters, Voltaire to Camus: "Micromegas" by Voltaire; "The Atheist's Mass" by Balzac; "Minuet" by de Maupassant; "The Guest" by Camus, six more. Excellent English translations on facing pages. Also French-English vocabulary list, exercises, more. 352pp. 5⅜ x 8½. 0-486-26443-2

CHICAGO AT THE TURN OF THE CENTURY IN PHOTOGRAPHS: 122 Historic Views from the Collections of the Chicago Historical Society, Larry A. Viskochil. Rare large-format prints offer detailed views of City Hall, State Street, the Loop, Hull House, Union Station, many other landmarks, circa 1904-1913. Introduction. Captions. Maps. 144pp. 9⅜ x 12¼. 0-486-24656-6

OLD BROOKLYN IN EARLY PHOTOGRAPHS, 1865-1929, William Lee Younger. Luna Park, Gravesend race track, construction of Grand Army Plaza, moving of Hotel Brighton, etc. 157 previously unpublished photographs. 165pp. 8⅞ x 11¾. 0-486-23587-4

THE MYTHS OF THE NORTH AMERICAN INDIANS, Lewis Spence. Rich anthology of the myths and legends of the Algonquins, Iroquois, Pawnees and Sioux, prefaced by an extensive historical and ethnological commentary. 36 illustrations. 480pp. 5⅜ x 8½. 0-486-25967-6

AN ENCYCLOPEDIA OF BATTLES: Accounts of Over 1,560 Battles from 1479 B.C. to the Present, David Eggenberger. Essential details of every major battle in recorded history from the first battle of Megiddo in 1479 B.C. to Grenada in 1984. List of Battle Maps. New Appendix covering the years 1967-1984. Index. 99 illustrations. 544pp. 6½ x 9¼. 0-486-24913-1

SAILING ALONE AROUND THE WORLD, Captain Joshua Slocum. First man to sail around the world, alone, in small boat. One of great feats of seamanship told in delightful manner. 67 illustrations. 294pp. 5⅜ x 8½. 0-486-20326-3

ANARCHISM AND OTHER ESSAYS, Emma Goldman. Powerful, penetrating, prophetic essays on direct action, role of minorities, prison reform, puritan hypocrisy, violence, etc. 271pp. 5⅜ x 8½. 0-486-22484-8

MYTHS OF THE HINDUS AND BUDDHISTS, Ananda K. Coomaraswamy and Sister Nivedita. Great stories of the epics; deeds of Krishna, Shiva, taken from puranas, Vedas, folk tales; etc. 32 illustrations. 400pp. 5⅜ x 8½. 0-486-21759-0

MY BONDAGE AND MY FREEDOM, Frederick Douglass. Born a slave, Douglass became outspoken force in antislavery movement. The best of Douglass' autobiographies. Graphic description of slave life. 464pp. 5⅜ x 8½. 0-486-22457-0

FOLLOWING THE EQUATOR: A Journey Around the World, Mark Twain. Fascinating humorous account of 1897 voyage to Hawaii, Australia, India, New Zealand, etc. Ironic, bemused reports on peoples, customs, climate, flora and fauna, politics, much more. 197 illustrations. 720pp. 5⅜ x 8½. 0-486-26113-1

THE PEOPLE CALLED SHAKERS, Edward D. Andrews. Definitive study of Shakers: origins, beliefs, practices, dances, social organization, furniture and crafts, etc. 33 illustrations. 351pp. 5⅜ x 8½. 0-486-21081-2

THE MYTHS OF GREECE AND ROME, H. A. Guerber. A classic of mythology, generously illustrated, long prized for its simple, graphic, accurate retelling of the principal myths of Greece and Rome, and for its commentary on their origins and significance. With 64 illustrations by Michelangelo, Raphael, Titian, Rubens, Canova, Bernini and others. 480pp. 5⅜ x 8½. 0-486-27584-1

PSYCHOLOGY OF MUSIC, Carl E. Seashore. Classic work discusses music as a medium from psychological viewpoint. Clear treatment of physical acoustics, auditory apparatus, sound perception, development of musical skills, nature of musical feeling, host of other topics. 88 figures. 408pp. 5⅜ x 8½. 0-486-21851-1

LIFE IN ANCIENT EGYPT, Adolf Erman. Fullest, most thorough, detailed older account with much not in more recent books, domestic life, religion, magic, medicine, commerce, much more. Many illustrations reproduce tomb paintings, carvings, hieroglyphs, etc. 597pp. 5⅜ x 8½. 0-486-22632-8

SUNDIALS, Their Theory and Construction, Albert Waugh. Far and away the best, most thorough coverage of ideas, mathematics concerned, types, construction, adjusting anywhere. Simple, nontechnical treatment allows even children to build several of these dials. Over 100 illustrations. 230pp. 5⅜ x 8½. 0-486-22947-5

THEORETICAL HYDRODYNAMICS, L. M. Milne-Thomson. Classic exposition of the mathematical theory of fluid motion, applicable to both hydrodynamics and aerodynamics. Over 600 exercises. 768pp. 6⅛ x 9¼. 0-486-68970-0

OLD-TIME VIGNETTES IN FULL COLOR, Carol Belanger Grafton (ed.). Over 390 charming, often sentimental illustrations, selected from archives of Victorian graphics–pretty women posing, children playing, food, flowers, kittens and puppies, smiling cherubs, birds and butterflies, much more. All copyright-free. 48pp. 9¼ x 12¼. 0-486-27269-9

PERSPECTIVE FOR ARTISTS, Rex Vicat Cole. Depth, perspective of sky and sea, shadows, much more, not usually covered. 391 diagrams, 81 reproductions of drawings and paintings. 279pp. 5⅜ x 8½. 0-486-22487-2

DRAWING THE LIVING FIGURE, Joseph Sheppard. Innovative approach to artistic anatomy focuses on specifics of surface anatomy, rather than muscles and bones. Over 170 drawings of live models in front, back and side views, and in widely varying poses. Accompanying diagrams. 177 illustrations. Introduction. Index. 144pp. 8⅜ x11¼. 0-486-26723-7

GOTHIC AND OLD ENGLISH ALPHABETS: 100 Complete Fonts, Dan X. Solo. Add power, elegance to posters, signs, other graphics with 100 stunning copyright-free alphabets: Blackstone, Dolbey, Germania, 97 more–including many lower-case, numerals, punctuation marks. 104pp. 8⅛ x 11. 0-486-24695-7

THE BOOK OF WOOD CARVING, Charles Marshall Sayers. Finest book for beginners discusses fundamentals and offers 34 designs. "Absolutely first rate . . . well thought out and well executed."–E. J. Tangerman. 118pp. 7¾ x 10⅝. 0-486-23654-4

ILLUSTRATED CATALOG OF CIVIL WAR MILITARY GOODS: Union Army Weapons, Insignia, Uniform Accessories, and Other Equipment, Schuyler, Hartley, and Graham. Rare, profusely illustrated 1846 catalog includes Union Army uniform and dress regulations, arms and ammunition, coats, insignia, flags, swords, rifles, etc. 226 illustrations. 160pp. 9 x 12. 0-486-24939-5

WOMEN'S FASHIONS OF THE EARLY 1900s: An Unabridged Republication of "New York Fashions, 1909," National Cloak & Suit Co. Rare catalog of mail-order fashions documents women's and children's clothing styles shortly after the turn of the century. Captions offer full descriptions, prices. Invaluable resource for fashion, costume historians. Approximately 725 illustrations. 128pp. 8⅜ x 11¼. 0-486-27276-1

HOW TO DO BEADWORK, Mary White. Fundamental book on craft from simple projects to five-bead chains and woven works. 106 illustrations. 142pp. 5⅜ x 8. 0-486-20697-1

THE 1912 AND 1915 GUSTAV STICKLEY FURNITURE CATALOGS, Gustav Stickley. With over 200 detailed illustrations and descriptions, these two catalogs are essential reading and reference materials and identification guides for Stickley furniture. Captions cite materials, dimensions and prices. 112pp. 6½ x 9¼. 0-486-26676-1

EARLY AMERICAN LOCOMOTIVES, John H. White, Jr. Finest locomotive engravings from early 19th century: historical (1804–74), main-line (after 1870), special, foreign, etc. 147 plates. 142pp. 11⅜ x 8¼. 0-486-22772-3

LITTLE BOOK OF EARLY AMERICAN CRAFTS AND TRADES, Peter Stockham (ed.). 1807 children's book explains crafts and trades: baker, hatter, cooper, potter, and many others. 23 copperplate illustrations. 140pp. 4⅝ x 6. 0-486-23336-7

VICTORIAN FASHIONS AND COSTUMES FROM HARPER'S BAZAR, 1867–1898, Stella Blum (ed.). Day costumes, evening wear, sports clothes, shoes, hats, other accessories in over 1,000 detailed engravings. 320pp. 9⅜ x 12¼. 0-486-22990-4

THE LONG ISLAND RAIL ROAD IN EARLY PHOTOGRAPHS, Ron Ziel. Over 220 rare photos, informative text document origin (1844) and development of rail service on Long Island. Vintage views of early trains, locomotives, stations, passengers, crews, much more. Captions. 8⅞ x 11¾. 0-486-26301-0

VOYAGE OF THE LIBERDADE, Joshua Slocum. Great 19th-century mariner's thrilling, first-hand account of the wreck of his ship off South America, the 35-foot boat he built from the wreckage, and its remarkable voyage home. 128pp. 5⅜ x 8½. 0-486-40022-0

TEN BOOKS ON ARCHITECTURE, Vitruvius. The most important book ever written on architecture. Early Roman aesthetics, technology, classical orders, site selection, all other aspects. Morgan translation. 331pp. 5⅜ x 8½. 0-486-20645-9

THE HUMAN FIGURE IN MOTION, Eadweard Muybridge. More than 4,500 stopped-action photos, in action series, showing undraped men, women, children jumping, lying down, throwing, sitting, wrestling, carrying, etc. 390pp. 7⅞ x 10⅝. 0-486-20204-6 Clothbd.

TREES OF THE EASTERN AND CENTRAL UNITED STATES AND CANADA, William M. Harlow. Best one-volume guide to 140 trees. Full descriptions, woodlore, range, etc. Over 600 illustrations. Handy size. 288pp. 4½ x 6⅜. 0-486-20395-6

GROWING AND USING HERBS AND SPICES, Milo Miloradovich. Versatile handbook provides all the information needed for cultivation and use of all the herbs and spices available in North America. 4 illustrations. Index. Glossary. 236pp. 5⅜ x 8½. 0-486-25058-X

BIG BOOK OF MAZES AND LABYRINTHS, Walter Shepherd. 50 mazes and labyrinths in all–classical, solid, ripple, and more–in one great volume. Perfect inexpensive puzzler for clever youngsters. Full solutions. 112pp. 8⅛ x 11. 0-486-22951-3

PIANO TUNING, J. Cree Fischer. Clearest, best book for beginner, amateur. Simple repairs, raising dropped notes, tuning by easy method of flattened fifths. No previous skills needed. 4 illustrations. 201pp. 5⅜ x 8½. 0-486-23267-0

HINTS TO SINGERS, Lillian Nordica. Selecting the right teacher, developing confidence, overcoming stage fright, and many other important skills receive thoughtful discussion in this indispensible guide, written by a world-famous diva of four decades' experience. 96pp. $5\frac{3}{8}$ x $8\frac{1}{2}$. 0-486-40094-8

THE COMPLETE NONSENSE OF EDWARD LEAR, Edward Lear. All nonsense limericks, zany alphabets, Owl and Pussycat, songs, nonsense botany, etc., illustrated by Lear. Total of 320pp. $5\frac{3}{8}$ x $8\frac{1}{2}$. (Available in U.S. only.) 0-486-20167-8

VICTORIAN PARLOUR POETRY: An Annotated Anthology, Michael R. Turner. 117 gems by Longfellow, Tennyson, Browning, many lesser-known poets. "The Village Blacksmith," "Curfew Must Not Ring Tonight," "Only a Baby Small," dozens more, often difficult to find elsewhere. Index of poets, titles, first lines. xxiii + 325pp. $5\frac{3}{8}$ x $8\frac{1}{4}$. 0-486-27044-0

DUBLINERS, James Joyce. Fifteen stories offer vivid, tightly focused observations of the lives of Dublin's poorer classes. At least one, "The Dead," is considered a masterpiece. Reprinted complete and unabridged from standard edition. 160pp. $5\frac{3}{16}$ x $8\frac{1}{4}$. 0-486-26870-5

GREAT WEIRD TALES: 14 Stories by Lovecraft, Blackwood, Machen and Others, S. T. Joshi (ed.). 14 spellbinding tales, including "The Sin Eater," by Fiona McLeod, "The Eye Above the Mantel," by Frank Belknap Long, as well as renowned works by R. H. Barlow, Lord Dunsany, Arthur Machen, W. C. Morrow and eight other masters of the genre. 256pp. $5\frac{3}{8}$ x $8\frac{1}{2}$. (Available in U.S. only.) 0-486-40436-6

THE BOOK OF THE SACRED MAGIC OF ABRAMELIN THE MAGE, translated by S. MacGregor Mathers. Medieval manuscript of ceremonial magic. Basic document in Aleister Crowley, Golden Dawn groups. 268pp. $5\frac{3}{8}$ x $8\frac{1}{2}$. 0-486-23211-5

THE BATTLES THAT CHANGED HISTORY, Fletcher Pratt. Eminent historian profiles 16 crucial conflicts, ancient to modern, that changed the course of civilization. 352pp. $5\frac{3}{8}$ x $8\frac{1}{2}$. 0-486-41129-X

NEW RUSSIAN-ENGLISH AND ENGLISH-RUSSIAN DICTIONARY, M. A. O'Brien. This is a remarkably handy Russian dictionary, containing a surprising amount of information, including over 70,000 entries. 366pp. $4\frac{1}{2}$ x $6\frac{1}{8}$. 0-486-20208-9

NEW YORK IN THE FORTIES, Andreas Feininger. 162 brilliant photographs by the well-known photographer, formerly with *Life* magazine. Commuters, shoppers, Times Square at night, much else from city at its peak. Captions by John von Hartz. 181pp. $9\frac{1}{4}$ x $10\frac{3}{4}$. 0-486-23585-8

INDIAN SIGN LANGUAGE, William Tomkins. Over 525 signs developed by Sioux and other tribes. Written instructions and diagrams. Also 290 pictographs. 111pp. $6\frac{1}{8}$ x $9\frac{1}{4}$. 0-486-22029-X

ANATOMY: A Complete Guide for Artists, Joseph Sheppard. A master of figure drawing shows artists how to render human anatomy convincingly. Over 460 illustrations. 224pp. $8\frac{3}{8}$ x $11\frac{1}{4}$. 0-486-27279-6

MEDIEVAL CALLIGRAPHY: Its History and Technique, Marc Drogin. Spirited history, comprehensive instruction manual covers 13 styles (ca. 4th century through 15th). Excellent photographs; directions for duplicating medieval techniques with modern tools. 224pp. $8\frac{3}{8}$ x $11\frac{1}{4}$. 0-486-26142-5

DRIED FLOWERS: How to Prepare Them, Sarah Whitlock and Martha Rankin. Complete instructions on how to use silica gel, meal and borax, perlite aggregate, sand and borax, glycerine and water to create attractive permanent flower arrangements. 12 illustrations. 32pp. 5⅜ x 8½. 0-486-21802-3

EASY-TO-MAKE BIRD FEEDERS FOR WOODWORKERS, Scott D. Campbell. Detailed, simple-to-use guide for designing, constructing, caring for and using feeders. Text, illustrations for 12 classic and contemporary designs. 96pp. 5⅜ x 8½. 0-486-25847-5

THE COMPLETE BOOK OF BIRDHOUSE CONSTRUCTION FOR WOODWORKERS, Scott D. Campbell. Detailed instructions, illustrations, tables. Also data on bird habitat and instinct patterns. Bibliography. 3 tables. 63 illustrations in 15 figures. 48pp. 5¼ x 8½. 0-486-24407-5

SCOTTISH WONDER TALES FROM MYTH AND LEGEND, Donald A. Mackenzie. 16 lively tales tell of giants rumbling down mountainsides, of a magic wand that turns stone pillars into warriors, of gods and goddesses, evil hags, powerful forces and more. 240pp. 5⅜ x 8½. 0-486-29677-6

THE HISTORY OF UNDERCLOTHES, C. Willett Cunnington and Phyllis Cunnington. Fascinating, well-documented survey covering six centuries of English undergarments, enhanced with over 100 illustrations: 12th-century laced-up bodice, footed long drawers (1795), 19th-century bustles, 19th-century corsets for men, Victorian "bust improvers," much more. 272pp. 5⅜ x 8¼. 0-486-27124-2

ARTS AND CRAFTS FURNITURE: The Complete Brooks Catalog of 1912, Brooks Manufacturing Co. Photos and detailed descriptions of more than 150 now very collectible furniture designs from the Arts and Crafts movement depict davenports, settees, buffets, desks, tables, chairs, bedsteads, dressers and more, all built of solid, quarter-sawed oak. Invaluable for students and enthusiasts of antiques, Americana and the decorative arts. 80pp. 6½ x 9¼. 0-486-27471-3

WILBUR AND ORVILLE: A Biography of the Wright Brothers, Fred Howard. Definitive, crisply written study tells the full story of the brothers' lives and work. A vividly written biography, unparalleled in scope and color, that also captures the spirit of an extraordinary era. 560pp. 6⅛ x 9¼. 0-486-40297-5

THE ARTS OF THE SAILOR: Knotting, Splicing and Ropework, Hervey Garrett Smith. Indispensable shipboard reference covers tools, basic knots and useful hitches; handsewing and canvas work, more. Over 100 illustrations. Delightful reading for sea lovers. 256pp. 5⅜ x 8½. 0-486-26440-8

FRANK LLOYD WRIGHT'S FALLINGWATER: The House and Its History, Second, Revised Edition, Donald Hoffmann. A total revision–both in text and illustrations–of the standard document on Fallingwater, the boldest, most personal architectural statement of Wright's mature years, updated with valuable new material from the recently opened Frank Lloyd Wright Archives. "Fascinating"–*The New York Times.* 116 illustrations. 128pp. 9¼ x 10¾. 0-486-27430-6

PHOTOGRAPHIC SKETCHBOOK OF THE CIVIL WAR, Alexander Gardner. 100 photos taken on field during the Civil War. Famous shots of Manassas Harper's Ferry, Lincoln, Richmond, slave pens, etc. 244pp. 10⅜ x 8¼. 0-486-22731-6

FIVE ACRES AND INDEPENDENCE, Maurice G. Kains. Great back-to-the-land classic explains basics of self-sufficient farming. The one book to get. 95 illustrations. 397pp. 5⅜ x 8½. 0-486-20974-1

THE MALLEUS MALEFICARUM OF KRAMER AND SPRENGER, translated by Montague Summers. Full text of most important witchhunter's "bible," used by both Catholics and Protestants. 278pp. 6⅝ x 10. 0-486-22802-9

SPANISH STORIES/CUENTOS ESPAÑOLES: A Dual-Language Book, Angel Flores (ed.). Unique format offers 13 great stories in Spanish by Cervantes, Borges, others. Faithful English translations on facing pages. 352pp. 5⅜ x 8½. 0-486-25399-6

GARDEN CITY, LONG ISLAND, IN EARLY PHOTOGRAPHS, 1869–1919, Mildred H. Smith. Handsome treasury of 118 vintage pictures, accompanied by carefully researched captions, document the Garden City Hotel fire (1899), the Vanderbilt Cup Race (1908), the first airmail flight departing from the Nassau Boulevard Aerodrome (1911), and much more. 96pp. 8⅞ x 11¾. 0-486-40669-5

OLD QUEENS, N.Y., IN EARLY PHOTOGRAPHS, Vincent F. Seyfried and William Asadorian. Over 160 rare photographs of Maspeth, Jamaica, Jackson Heights, and other areas. Vintage views of DeWitt Clinton mansion, 1939 World's Fair and more. Captions. 192pp. 8⅞ x 11. 0-486-26358-4

CAPTURED BY THE INDIANS: 15 Firsthand Accounts, 1750-1870, Frederick Drimmer. Astounding true historical accounts of grisly torture, bloody conflicts, relentless pursuits, miraculous escapes and more, by people who lived to tell the tale. 384pp. 5⅜ x 8½. 0-486-24901-8

THE WORLD'S GREAT SPEECHES (Fourth Enlarged Edition), Lewis Copeland, Lawrence W. Lamm, and Stephen J. McKenna. Nearly 300 speeches provide public speakers with a wealth of updated quotes and inspiration–from Pericles' funeral oration and William Jennings Bryan's "Cross of Gold Speech" to Malcolm X's powerful words on the Black Revolution and Earl of Spenser's tribute to his sister, Diana, Princess of Wales. 944pp. 5⅜ x 8⅜. 0-486-40903-1

THE BOOK OF THE SWORD, Sir Richard F. Burton. Great Victorian scholar/adventurer's eloquent, erudite history of the "queen of weapons"–from prehistory to early Roman Empire. Evolution and development of early swords, variations (sabre, broadsword, cutlass, scimitar, etc.), much more. 336pp. 6⅛ x 9¼. 0-486-25434-8

AUTOBIOGRAPHY: The Story of My Experiments with Truth, Mohandas K. Gandhi. Boyhood, legal studies, purification, the growth of the Satyagraha (nonviolent protest) movement. Critical, inspiring work of the man responsible for the freedom of India. 480pp. 5⅜ x 8½. (Available in U.S. only.) 0-486-24593-4

CELTIC MYTHS AND LEGENDS, T. W. Rolleston. Masterful retelling of Irish and Welsh stories and tales. Cuchulain, King Arthur, Deirdre, the Grail, many more. First paperback edition. 58 full-page illustrations. 512pp. 5⅜ x 8½. 0-486-26507-2

THE PRINCIPLES OF PSYCHOLOGY, William James. Famous long course complete, unabridged. Stream of thought, time perception, memory, experimental methods; great work decades ahead of its time. 94 figures. 1,391pp. 5⅜ x 8½. 2-vol. set.
Vol. I: 0-486-20381-6 Vol. II: 0-486-20382-4

THE WORLD AS WILL AND REPRESENTATION, Arthur Schopenhauer. Definitive English translation of Schopenhauer's life work, correcting more than 1,000 errors, omissions in earlier translations. Translated by E. F. J. Payne. Total of 1,269pp. 5⅜ x 8½. 2-vol. set. Vol. 1: 0-486-21761-2 Vol. 2: 0-486-21762-0

MAGIC AND MYSTERY IN TIBET, Madame Alexandra David-Neel. Experiences among lamas, magicians, sages, sorcerers, Bonpa wizards. A true psychic discovery. 32 illustrations. 321pp. 5⅜ x 8½. (Available in U.S. only.) 0-486-22682-4

THE EGYPTIAN BOOK OF THE DEAD, E. A. Wallis Budge. Complete reproduction of Ani's papyrus, finest ever found. Full hieroglyphic text, interlinear transliteration, word-for-word translation, smooth translation. 533pp. 6½ x 9¼. 0-486-21866-X

HISTORIC COSTUME IN PICTURES, Braun & Schneider. Over 1,450 costumed figures in clearly detailed engravings–from dawn of civilization to end of 19th century. Captions. Many folk costumes. 256pp. 8⅜ x 11¾. 0-486-23150-X

MATHEMATICS FOR THE NONMATHEMATICIAN, Morris Kline. Detailed, college-level treatment of mathematics in cultural and historical context, with numerous exercises. Recommended Reading Lists. Tables. Numerous figures. 641pp. 5⅜ x 8½. 0-486-24823-2

PROBABILISTIC METHODS IN THE THEORY OF STRUCTURES, Isaac Elishakoff. Well-written introduction covers the elements of the theory of probability from two or more random variables, the reliability of such multivariable structures, the theory of random function, Monte Carlo methods of treating problems incapable of exact solution, and more. Examples. 502pp. 5⅜ x 8½. 0-486-40691-1

THE RIME OF THE ANCIENT MARINER, Gustave Doré, S. T. Coleridge. Doré's finest work; 34 plates capture moods, subtleties of poem. Flawless full-size reproductions printed on facing pages with authoritative text of poem. "Beautiful. Simply beautiful."–*Publisher's Weekly*. 77pp. 9¼ x 12. 0-486-22305-1

SCULPTURE: Principles and Practice, Louis Slobodkin. Step-by-step approach to clay, plaster, metals, stone; classical and modern. 253 drawings, photos. 255pp. 8⅛ x 11. 0-486-22960-2

THE INFLUENCE OF SEA POWER UPON HISTORY, 1660–1783, A. T. Mahan. Influential classic of naval history and tactics still used as text in war colleges. First paperback edition. 4 maps. 24 battle plans. 640pp. 5⅜ x 8½. 0-486-25509-3

THE STORY OF THE TITANIC AS TOLD BY ITS SURVIVORS, Jack Winocour (ed.). What it was really like. Panic, despair, shocking inefficiency, and a little heroism. More thrilling than any fictional account. 26 illustrations. 320pp. 5⅜ x 8½. 0-486-20610-6

ONE TWO THREE . . . INFINITY: Facts and Speculations of Science, George Gamow. Great physicist's fascinating, readable overview of contemporary science: number theory, relativity, fourth dimension, entropy, genes, atomic structure, much more. 128 illustrations. Index. 352pp. 5⅜ x 8½. 0-486-25664-2

DALÍ ON MODERN ART: The Cuckolds of Antiquated Modern Art, Salvador Dalí. Influential painter skewers modern art and its practitioners. Outrageous evaluations of Picasso, Cézanne, Turner, more. 15 renderings of paintings discussed. 44 calligraphic decorations by Dalí. 96pp. 5⅜ x 8½. (Available in U.S. only.) 0-486-29220-7

ANTIQUE PLAYING CARDS: A Pictorial History, Henry René D'Allemagne. Over 900 elaborate, decorative images from rare playing cards (14th–20th centuries): Bacchus, death, dancing dogs, hunting scenes, royal coats of arms, players cheating, much more. 96pp. 9¼ x 12¼. 0-486-29265-7

MAKING FURNITURE MASTERPIECES: 30 Projects with Measured Drawings, Franklin H. Gottshall. Step-by-step instructions, illustrations for constructing handsome, useful pieces, among them a Sheraton desk, Chippendale chair, Spanish desk, Queen Anne table and a William and Mary dressing mirror. 224pp. 8⅛ x 11¼. 0-486-29338-6

NORTH AMERICAN INDIAN DESIGNS FOR ARTISTS AND CRAFTSPEOPLE, Eva Wilson. Over 360 authentic copyright-free designs adapted from Navajo blankets, Hopi pottery, Sioux buffalo hides, more. Geometrics, symbolic figures, plant and animal motifs, etc. 128pp. 8⅜ x 11. (Not for sale in the United Kingdom.) 0-486-25341-4

THE FOSSIL BOOK: A Record of Prehistoric Life, Patricia V. Rich et al. Profusely illustrated definitive guide covers everything from single-celled organisms and dinosaurs to birds and mammals and the interplay between climate and man. Over 1,500 illustrations. 760pp. 7½ x 10⅛. 0-486-29371-8

VICTORIAN ARCHITECTURAL DETAILS: Designs for Over 700 Stairs, Mantels, Doors, Windows, Cornices, Porches, and Other Decorative Elements, A. J. Bicknell & Company. Everything from dormer windows and piazzas to balconies and gable ornaments. Also includes elevations and floor plans for handsome, private residences and commercial structures. 80pp. 9⅜ x 12¼. 0-486-44015-X

WESTERN ISLAMIC ARCHITECTURE: A Concise Introduction, John D. Hoag. Profusely illustrated critical appraisal compares and contrasts Islamic mosques and palaces–from Spain and Egypt to other areas in the Middle East. 139 illustrations. 128pp. 6 x 9. 0-486-43760-4

CHINESE ARCHITECTURE: A Pictorial History, Liang Ssu-ch'eng. More than 240 rare photographs and drawings depict temples, pagodas, tombs, bridges, and imperial palaces comprising much of China's architectural heritage. 152 halftones, 94 diagrams. 232pp. 10¾ x 9⅞. 0-486-43999-2

THE RENAISSANCE: Studies in Art and Poetry, Walter Pater. One of the most talked-about books of the 19th century, *The Renaissance* combines scholarship and philosophy in an innovative work of cultural criticism that examines the achievements of Botticelli, Leonardo, Michelangelo, and other artists. "The holy writ of beauty."–Oscar Wilde. 160pp. 5⅜ x 8½. 0-486-44025-7

A TREATISE ON PAINTING, Leonardo da Vinci. The great Renaissance artist's practical advice on drawing and painting techniques covers anatomy, perspective, composition, light and shadow, and color. A classic of art instruction, it features 48 drawings by Nicholas Poussin and Leon Battista Alberti. 192pp. 5⅜ x 8½. 0-486-44155-5

THE MIND OF LEONARDO DA VINCI, Edward McCurdy. More than just a biography, this classic study by a distinguished historian draws upon Leonardo's extensive writings to offer numerous demonstrations of the Renaissance master's achievements, not only in sculpture and painting, but also in music, engineering, and even experimental aviation. 384pp. 5⅜ x 8½. 0-486-44142-3

WASHINGTON IRVING'S RIP VAN WINKLE, Illustrated by Arthur Rackham. Lovely prints that established artist as a leading illustrator of the time and forever etched into the popular imagination a classic of Catskill lore. 51 full-color plates. 80pp. 8⅜ x 11. 0-486-44242-X

HENSCHE ON PAINTING, John W. Robichaux. Basic painting philosophy and methodology of a great teacher, as expounded in his famous classes and workshops on Cape Cod. 7 illustrations in color on covers. 80pp. 5⅜ x 8½. 0-486-43728-0